CW00735748

ON TRIAL FOR REASON

ON TRIAL *for* REASON

Science, Religion, and Culture
in the Galileo Affair

MAURICE A. FINOCCHIARO

OXFORD
UNIVERSITY PRESS

Great Clarendon Street, Oxford, OX2 6DP,
United Kingdom

Oxford University Press is a department of the University of Oxford.
It furthers the University's objective of excellence in research, scholarship,
and education by publishing worldwide. Oxford is a registered trade mark of
Oxford University Press in the UK and in certain other countries

First Edition published in 2019

Impression: 1

Published in the United States of America by Oxford University Press
198 Madison Avenue, New York, NY 10016, United States of America

British Library Cataloguing in Publication Data
Data available

Library of Congress Control Number: 2018968360

ISBN 978–0–19–879792–0

Printed and bound in Great Britain by
Clays Ltd, Elcograf S.p.A.

PREFACE AND ACKNOWLEDGMENTS

This book is meant to be a summary, synthesis, and simplification of my many scholarly works dealing with the Galileo affair: his Inquisition trial, its intellectual issues, its background, the historical aftermath up to our day, and the philosophical and cultural lessons involving the relationship between science and religion and the nature of rationality, scientific method, and critical thinking. Thus, I owe a debt of gratitude to the many scholars and institutions from whom my scholarly work has benefitted. They are mentioned in my previous works, and so here it will suffice for me to express only this general acknowledgment.

Furthermore, this book is also meant to be an expansion and elaboration of a public lecture which I have had the opportunity to deliver at many venues in many parts of the world. In fact, after I started publishing scholarly works on the Galileo affair, it did not take long before I received invitations to present a one-hour lecture to audiences of intelligent and educated persons who were not specialists on the topic, but were curious and interested about its details. Although initially challenging and uncomfortable, such lecturing became increasingly pleasant, and also beneficial to me by forcing me to focus on the universal and perennial relevance of the topic. I have never had the occasion to formally thank the organizers and audiences of such lectures, and so here it seems very appropriate to express my gratitude, both generally and specifically.

Deserving mention are the following more memorable occasions, when the organizers were especially wise and effective, and the audiences especially engaged and engaging: Raymond Erickson and the audience of musicians, musicologists, artists, and art critics, at the Tenth Aston Magna Academy on Music, the Arts, and Society, on the theme of "Foundations of the Italian Baroque, 1560–1620," at Rutgers State University, June 23–30,

1991; Carlos Alvarez and the interdisciplinary audience of faculty and students, at the Colloquium on "Mathematics, History, and Culture," to inaugurate the new library of the Faculty of Science, at the National Autonomous University of Mexico (UNAM), Mexico City, September 20–2, 1994; Derrick Pitts and the general public at "The Legacy of Galileo Symposium," on the occasion of the International Year of Astronomy, to commemorate the 400th anniversary of Galileo's telescopic discoveries, at the Franklin Institute, Philadelphia, June 18–20, 2009; Peter Slezak and the audience at the public lecture sponsored by the Program in History and Philosophy of Science and the Faculty of Arts and Social Sciences, University of New South Wales, Sydney, Australia, also to celebrate the International Year of Astronomy, October 22, 2009; John W. Meriwether and the audience of scientists, engineers, and spouses, at the "Classmate Speaker Program," 50th Anniversary Reunion of the Class of 1964, Massachusetts Institute of Technology, June 6, 2014; and Kenneth Wolfe and the audience of mostly elderly and retired persons at the "Contextual Lecture Series 2014: People Who Changed the World," Dulwich Picture Gallery, London, July 8, 2014.

Last but not least, there is a group of persons to whom I am grateful and who deserve acknowledgment for their assistance concerning this particular book, attempting to simplify my specialized scholarship and to amplify my public lecture. John Heilbron, author of *Galileo* and many other books in the history of physics, provided both positive support and substantive criticism; I appreciated both, while feeling free to criticize his criticism. Latha Menon, at Oxford University Press, not only did the usual tasks required of a commissioning editor, but also read and edited the entire manuscript; her suggestions significantly improved the literary style and narrative and the general appeal of my writing. One of the anonymous reviewers of my original proposal provided not only a favorable recommendation, but also a very perceptive and fruitful interpretation of my project: "without 'dumbing down' of historical and philosophical content, to be of interest to a wide readership"; this was and continued to be my guiding principle in the writing of this book, and I found this reviewer's judgment a constant source of encouragement. Finally, two readers of my manuscript deserve special thanks: Aaron Abbey, a professional librarian at the University of Nevada, Las Vegas and a former student of mine; and Mark Attorri, a professional and practicing attorney in New Hampshire, for whom Galileo is a hobby and diversion from his law practice. They provided numerous and detailed comments and criticisms, and I feel extremely fortunate to have had this book read by these two ideal examples of intelligent, educated, interested, curious, and nonspecialist readers.

In a class by itself is the special debt of gratitude which I owe to the University of Nevada, Las Vegas, in particular the College of Liberal Arts, the Philosophy Department, the Chair of the Philosophy Department (David Forman), my departmental colleagues (David Beisecker, Ian Dove, Todd Jones, Bill Ramsey, Paul Schollmeier, and James Woodbridge), and the staff in the Office of Information Technology (Hector Ibarra and Nick Panissidi). They have continued to provide institutional, material, and moral support, even after I decided to retire from formal teaching to work full time on research, scholarship, and writing.

CONTENTS

CHAPTER 1

INTRODUCTION

Avoiding Myths and Muddles

Galileo's Legacy

In 1633, at the conclusion of one of history's most famous trials, the Roman Inquisition found Galileo Galilei guilty of "vehement suspicion of heresy"; this was a specific category of religious crime intermediate in seriousness between formal heresy and mild suspicion of heresy. He had committed this alleged crime by defending the idea that the Earth is a planet rotating daily around its own axis and revolving yearly around the Sun; his argument was found in a book published the previous year and entitled *Dialogue on the Two Chief World Systems, Ptolemaic and Copernican*. The problem stemmed chiefly from the fact that Galileo was implicitly denying the Catholic Church's beliefs that the Earth's motion contradicted Scripture and Scripture was a scientific authority.

Thus, Galileo became the protagonist of a cause célèbre that continues to our own day. For example, in the eighteenth century, Voltaire opined that the tragedy would bring "eternal disgrace" to the Catholic Church; and in the twentieth century, Arthur Koestler labeled it "the greatest scandal in Christendom."[1]

However, there is also irony in this tragedy. For, as we shall see later, eventually the Church came to recognize that Galileo was right not only about the Earth's motion, but also about the limited authority of Scripture. This recognition came in 1893 when Pope Leo XIII issued an encyclical entitled *Providentissimus Deus*, propounding the Galilean principle that Scripture is not a scientific authority, but only one on questions of faith and morals. Moreover, another acknowledgment came in the period 1979–92, when Pope Saint John Paul II undertook a highly publicized and highly controversial "rehabilitation" of Galileo. John Paul made it clear and explicit that Galileo had been theologically right about biblical hermeneutics, as against his ecclesiastical opponents; moreover, the pope credited Galileo with having preached, practiced, and embodied the very important principle that religion and science are really in harmony, and not incompatible. In short, the world's oldest religious institution, which continues to be one of the world's great religions, has found ways and reasons to try to appropriate Galileo's legacy with regard to two basic principles, involving the limited authority of Scripture and the harmonious relationship between science and religion.

It is not surprising that the Catholic Church would try to appropriate Galileo's legacy. In fact, independently of his epoch-making role in the history and philosophy of religion, his legacy has a second main aspect: Galileo was one of the founders of modern science. That is, science as we know it today emerged in the sixteenth and seventeenth centuries thanks to the discoveries, inventions, ideas, and activities of a group of people like Galileo that also included Nicolaus Copernicus, Johannes Kepler, René Descartes, Christiaan Huygens, and Isaac Newton. Frequently, Galileo is singled out as the most pivotal of these founders and called the Father of Modern Science. Although many people have repeated or elaborated such a characterization, it is significant that it originates in the judgment of practicing scientists themselves, such as Albert Einstein and Stephen Hawking.[2] Galileo's most important contributions involved physics, astronomy, and scientific method.

In physics, Galileo pioneered the experimental investigation of motion. He also formulated, clarified, and systematized many of the basic concepts and principles needed for the theoretical analysis of motion, such as an approximation to the law of inertia, a formulation of the relativity of motion, and the composition of motion into distinct components. And he discovered the laws of falling bodies, including free fall, descent on inclined planes, pendulums, and projectiles.

In astronomy, Galileo introduced the telescope as an instrument for systematic observation. He made a number of crucial observational discoveries, such as mountains of the Moon, satellites of Jupiter, phases of Venus, and sunspots. And he understood the cosmological significance of these observational facts and gave essentially correct interpretations of many of them; that is, he provided a robust confirmation of the theory that the Earth moves, daily around its own axis and yearly around the Sun.

With regard to scientific method, Galileo pioneered several important practices. For example, he was a leader in the use of artificial instruments (like the telescope) to learn new facts about the world; this is to be contrasted to the use of instruments like the compass for practical purposes. Moreover, he pioneered the active intervention into and exploratory manipulation of physical phenomena in order to gain access to aspects of nature that are not detectable without such experimentation; this is the essence of the experimental method, as distinct from a merely observational approach. He also contributed to the establishment and extension of other more traditional, but little used, methodological practices, such as the use of a quantitative and mathematical approach in the study of motion. He contributed to the explicit formulation and clarification of important methodological principles, such as the setting aside of biblical assertions and religious authority in scientific inquiry. And he was also an inventor, making significant contributions to the devising and improvement of such instruments as the telescope, microscope, thermometer, and pendulum clock.

Finally, there is a third aspect to Galileo's legacy. In fact, the historical circumstances of his time and his own personal inclinations made him into a kind of philosopher. Of course, he was not a systematic metaphysician who speculated about the eternal problems of being and nothingness. Instead he was a concrete-oriented and practical-oriented critical thinker who not only was engaged in a quest for knowledge of nature, but also reflected on questions about the nature of knowledge. In the eloquent words of Owen Gingerich, for Galileo "what was at issue was both the truth of nature and the nature of truth."[3] Now, let us define *epistemology* as the branch of philosophy that studies the nature of knowledge in general, and scientific knowledge in particular, including the principles and procedures that are useful in the acquisition of knowledge; if one focuses on just the latter principles and procedures, that defines the branch of epistemology called *methodology*. In this regard, Galileo was like the ancient Greek philosopher Socrates, their main difference being that Socrates reflected on moral or ethical questions of good and evil and the meaning of life. Thus, just as many regard Socrates as the Father of Western Philosophy, we may regard Galileo as the Socrates of methodology and epistemology.

That is, as already hinted at and as we shall see in more detail later, Galileo's contributions to scientific knowledge were so radical that he constantly had to discuss with his opponents (scientific as well as ecclesiastic) not only what were the observational facts and what was their best theoretical interpretation, but also what were the proper rules for establishing the facts and for interpreting them. With scientific opponents he had to discuss questions such as whether artificial instruments like the telescope have a legitimate role in learning new truths about reality; whether scientific authorities, such as Aristotle (384–322 BC), should be relied upon to the exclusion of one's own independent judgment; whether mathematics has an important, and perhaps essential, role to play in the study of natural phenomena; and so on. With ecclesiastic opponents, Galileo had to discuss whether Scripture should be treated as a source of scientific information about physical reality; whether scientific theories that contradict the literal meaning of Scripture should be summarily rejected or treated as hypotheses; whether hypotheses are potentially true descriptions of reality or merely convenient instruments of calculation and prediction; and so on.

In short, Galileo's legacy clearly has a three-fold character, relating to science, religion, and philosophy. These three things are such major and crucial cultural elements, and their interaction has such significant cultural ramifications, that we may also speak more generally of his cultural legacy.

In this book, all aspects of Galileo's cultural legacy are discussed by focusing on his trial by the Inquisition, stressing its intellectual developments and issues, and elaborating, in turn, its background, proceedings, aftermath, and significance. However, before articulating the background, it is important to have a methodological discussion outlining the multifaceted and balanced approach that is necessary to avoid common pitfalls. This approach requires a mastery of a number of distinctions, which, however, must not be turned into separations. It also requires an awareness of the non-intellectual factors, which cannot be totally neglected, even as one stresses intellectual aspects. Finally, this approach requires an awareness of such pedestrian things as dates, places, and names, and so a rudimentary sketch of Galileo's life will be provided presently; such a biographical overview can also serve as a synthetic preview of this book's chapters. To these three topics, I now turn.

Biographical Highlights

Galileo was born in Pisa in 1564 (Figure 1). At that time, the northern Italian city of Pisa was part of an independent state known as the Grand Duchy of Tuscany, which had Florence as its capital, and which was ruled by a grand duke of the famous House of the Medici. Pisa already had the claim to fame which it continues to have today: it was the site of a massive and monumental tower that had been leaning without falling ever since it was built in the thirteenth century. Thus, some myth makers have invented the legend that at some point Galileo performed experiments by dropping two cannon balls of different weights from the top of the tower, to test whether they reached the ground at the same time. Similarly, the year of Galileo's birth happened to be the same as that of Michelangelo's death, and this coincidence has led other myth makers to speak of a transmigration of a specially gifted soul from Michelangelo to Galileo. However, this particular fable overlooks the possibility that such a talented soul might have migrated to Shakespeare, who was also born in 1564.

Be that as it may, Galileo's father, Vincenzo, was a musician, composer, and musicologist. Vincenzo was highly skillful at playing the lute, and a prolific writer and arranger of songs, but he was and remains best known for his revolutionary theories of music. In this field of musicology, his most

Figure 1. Portrait of Galileo, published in his *Sunspots* (1613) and *Assayer* (1623)

relevant accomplishment was to preach and practice the need to test experimentally the empirical accuracy of rules of harmony. His many experiments dealt with how the audible and melodious sound produced by plucking a string changed as a result of various changes in the properties of the string; that is, changes in its length, tension, weight, material, etc. Such experimentation had not been done or updated in any sustained manner for two thousand years, since the ancient Greeks.

Vincenzo's innovation is important because there can be no doubt that it was from his father that Galileo learned to practice and to appreciate the experimental method. However, Galileo also had his own original and revolutionary idea—utilizing experimentation in the study of falling bodies. Thus, Galileo's own experiments dealt with how the speed of a falling body changed as a result of various changes in the parameters of the motion; that is, changes in height, time, weight, substance, impediments, etc.

In 1581, Galileo enrolled at the University of Pisa to study medicine, but soon switched to mathematics, which he also studied privately outside the university. After four years, he left the university without a degree and began to do private teaching and independent research. In 1589, he obtained a position as professor of mathematics at the University of Pisa, and then from 1592 to 1610 at the University of Padua.

At that time, Padua was part of the Republic of Venice, which was an independent state that included some territory in north-eastern Italy, some parts of the eastern coast of the Adriatic Sea, and some islands in the south-eastern Mediterranean Sea. The republic was then at the height of its power, despite continuing skirmishes and conflicts with the Ottoman Empire. In fact, Venice had led the European coalition that defeated the Ottoman Turks in the great naval battle of Lepanto in 1571. Thus, the city of Venice was at that time a world center of commerce and wealth, and of arts and culture.

This is important because the University of Padua was a state institution, and so for 18 years Galileo was a Venetian state employee. Moreover, Padua was located only 26 miles away from the city of Venice, and so during those 18 years Galileo often went there for business, entertainment, socializing, etc. Thus, it is not surprising that from a personal point of view those years were one of the happiest periods of his life. Still, he missed his native Tuscany and always regarded himself as a Tuscan citizen; he never quite accepted life as a Venetian, and did all he could to acquire a position in his native region.

Professionally speaking, during this Paduan period, Galileo was researching primarily the nature of motion. He was critical of the physics of Aristotle, and favorably inclined toward the statics and mathematics of another ancient Greek, Archimedes of Syracuse (287–212 BC). Galileo made great progress in this project, and it was then that he made most of his contributions and discoveries in physics. However, he did not publish any of these results during this earlier period.

Moreover, Galileo was acquainted with the theory of a moving Earth advanced by Nicolaus Copernicus (1473–1543) in his book *On the Revolutions of the Heavenly Spheres* (1543) (Figure 2). He was appreciative of the fact that Copernicus had advanced a novel argument. Galileo had also intuited that the Earth's motion was more consistent in general with the new physics he was then developing than was the Earth standing still at the center of the universe; in particular, he had been attracted to Copernicanism because he felt that the Earth's motion could best explain why the tides occur. However, he was acutely aware of the considerable evidence against Copernicanism, especially that stemming from astronomical observation; for example, the failure to detect annual changes in the fixed stars, to see phases for the planet Venus, and to discern similarities between the Earth and the heavenly bodies. Thus, Galileo judged that the anti-Copernican arguments far outweighed the pro-Copernican ones.

However, his telescopic discoveries led Galileo to a major re-assessment of Copernicanism. In 1609, he perfected the telescope to such an extent as to make it an astronomically useful instrument that could not be duplicated by others for some time. By its means he made several startling discoveries which he immediately published in *The Sidereal Messenger* (1610): mountains on the Moon, satellites around the planet Jupiter, stellar composition of the

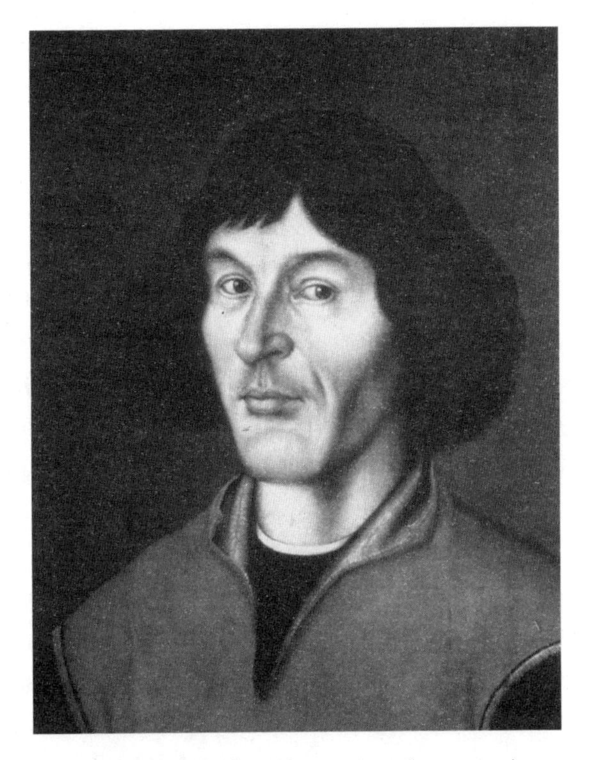

Figure 2. Nicolaus Copernicus (1473–1543)

Milky Way and nebulas, and countless stars never seen before. As a result, he became a celebrity, resigned his professorship at Padua, was appointed Philosopher and Chief Mathematician to the grand duke of Tuscany, and moved to Florence the same year. Soon thereafter, he also discovered the phases of Venus and sunspots; on the latter, he published the *History and Demonstrations Concerning Sunspots* (1613). Although he realized that these discoveries did not conclusively establish the Copernican theory, he had no doubt that they confirmed it.

This realization also encouraged Galileo to be more critical of the theological objections to Copernicanism, although he was careful and prudent because he knew that this aspect of the problem was potentially dangerous and explosive. His telescopic discoveries and their Copernican interpretation had immediately been criticized on scriptural grounds. At first, he ignored such attacks. But eventually he felt he could not remain silent, and decided to refute the biblical argument against Copernicus. To avoid scandalous publicity, he wrote his criticism in the form of long private letters, in December 1613 to his disciple Benedetto Castelli, and in spring 1615 to the grand duchess dowager Christina.

Galileo's letters circulated widely and the conservatives got even more upset. Thus, in February 1615, a Dominican friar filed a written complaint against Galileo with the Inquisition in Rome. An investigation was launched that lasted about a year. As part of this inquiry, a committee of Inquisition consultants reported that the key Copernican theses about the Earth's motion were absurd and false in natural philosophy and heretical in theology. The Inquisition also interrogated other witnesses.

However, Galileo himself was not summoned or interrogated, partly because the key witnesses exonerated him and partly because Galileo's letters had not been published, whereas his published writings contained neither a categorical assertion of Copernicanism nor a denial of the scientific authority of Scripture. Nor did the Inquisition issue a formal condemnation of Copernicanism as a heresy. Instead two milder consequences followed.

In February 1616, Galileo himself was given a private warning by Cardinal-Inquisitor Robert Bellarmine forbidding him to hold or defend the truth of the Earth's motion; Galileo agreed to comply. And in March, there was a decree by the Congregation of the Index, the less authoritative department in charge of book censorship: without mentioning Galileo at all, it declared the Copernican doctrine false and contrary to Scripture, and it temporarily banned Copernicus's 1543 book.

For the next several years, Galileo kept quiet about the forbidden topic, until 1623 when Florentine Cardinal Maffeo Barberini became Pope Urban VIII. Barberini was an old admirer and patron, and so Galileo felt freer and decided to write the book on the system of the world conceived earlier, adapting its form to the new restrictions. Galileo wrote the book in the form of a dialogue among three characters engaged in a critical discussion of the cosmological, astronomical, physical, and philosophical arguments, but avoiding the biblical or theological ones.

This *Dialogue on the Two Chief World Systems, Ptolemaic and Copernican* was published in 1632. Its key thesis is that the arguments favoring the Earth's motion were stronger than those favoring the Earth's rest, and in that sense Copernicanism was more probable than the traditional view. Galileo managed to incorporate into the discussion the new telescopic discoveries, his conclusions about the physics of moving bodies, frequent methodological reflections, and an explanation of the tides in terms of the Earth's motion (the so-called tidal argument).

However, his enemies raised several complaints. A key charge was that the book defended the truth of the Earth's motion, which he had been forbidden to do. Thus, he was summoned to Rome to stand trial, which began in April 1633. At the first interrogation, Galileo denied that his book defended the Earth's motion, claiming instead that it was a critical examination of the arguments on both sides. There followed an out-of-court plea-bargaining meeting, during which he was persuaded to plead guilty to this charge in exchange for leniency. So, at the next deposition, he admitted having defended the Earth's motion, but insisted that this was unintentional. The trial concluded with a sentence that found him guilty of "vehement suspicion of heresy," an intermediate category of religious crime; the *Dialogue* was banned; and he was to be under indefinite house arrest.

One of the ironic results of this condemnation was that, after the trial, to keep his sanity, Galileo went back to his earlier research on motion. Thus, he organized his old notes, and five years later published his most important contribution to physics, the *Two New Sciences* (1638). Without the tragedy of the trial, he might have never done it. He died in Arcetri (near Florence) in 1642, surrounded by family and disciples.

Galileo's condemnation started a controversy that shows no signs of abating. He became a cultural icon. His tragedy acquired paradigmatic significance for the perennial and universal problem of the relationship between science and religion. And his manner of thinking and search for truth became a model of rationality, scientific method, and critical thinking, to be considered by anyone engaged in such a search.

A New Approach to Galileo's Trial

The most common view about the trial of Galileo is that it epitomizes the conflict between enlightened science and obscurantist religion. One version of this view may be gathered from an inscription on a public monument in Rome near Villa Medici. This is the palace where Galileo resided on some of his visits to Rome, and where he was held under house arrest for about a week after the 1633 sentence. The inscription reads: "The palace next to this spot, / which belonged formerly to the Medici, / was a prison for Galileo Galilei, / guilty of having seen / the earth turn around the sun."

The historical and cultural importance of this minor tourist attraction is that it expresses some of the most common myths widely held about the trial of Galileo: that, with his telescope, he "saw" the Earth's motion (an observation still impossible to make even in the twenty-first century); that he was "imprisoned" by the Inquisition (whereas he was held under house arrest); and that his crime was to have discovered the truth. Now, since to condemn someone for this reason can result only from ignorance, prejudice, and narrow-mindedness, we also have here a formulation of the myth that alleges the incompatibility between science and religion.

The incompatibility thesis is very widespread. For example, various formulations of it have been advanced by such scientific, philosophical, and cultural icons as Voltaire, Bertrand Russell, Albert Einstein, and Karl Popper. However, I believe that such a thesis is erroneous, misleading, and simplistic, and will return to consider it in depth in Chapter 9.

For the moment, one main reason for identifying this first anti-clerical myth about the trial is that it may be usefully contrasted to a second myth at the opposite extreme. It seems that some found it appropriate to fight an objectionable myth by constructing another.

The opposite anti-Galilean myth maintains that Galileo deserved condemnation because he violated not only various ecclesiastical norms, but also various rules of scientific methodology and logical reasoning; he is thus portrayed as a master of cunning and knavery, and it is difficult to find a misdeed of which the proponents of this myth have not accused him. The history of this myth too has its own fascination; it too includes illustrious names, such as French physicist, philosopher, and historian Pierre Duhem, German playwright Bertolt Brecht, Hungarian intellectual Arthur Koestler, and Austrian-American philosopher Paul Feyerabend; and this myth and its underlying thesis will also be discussed in more detail later (Chapters 8 and 9).

These two opposite myths are useful as reference points in order to orient oneself in the study of the controversy, since it is impossible to evaluate the trial adequately unless one admits that both of these accounts are mythological and thus rejects both. However, avoiding them is easier said than done. For example, one cannot simply follow a mechanical approach of mediating a compromise by dividing in half the difference that separates them. A helpful way of proceeding is to read the relevant texts and documents with care and with an awareness of a number of crucial conceptual distinctions.

One of the most important of these distinctions is that the trial of Galileo involved both questions about the truth of nature and the nature of truth, to use Owen Gingerich's (1982) eloquent expression quoted earlier. That is, the controversy was at least two-sided: it involved partly *scientific issues* about physical facts, natural phenomena, and astronomical and cosmological matters; and it also involved *methodological* and *epistemological questions* about what truth is and the proper way to search for it, and about what knowledge is and how to acquire it.

The overarching scientific issue was whether the Earth stands still at the center of the universe, with all heavenly bodies revolving around it, or whether the Earth is itself a heavenly body that rotates on its axis every day and revolves around the Sun once a year. There were several distinct but interrelated questions here. One was whether the whole universe revolves daily from east to west around a motionless Earth, or the Earth alone rotates daily on its axis in the opposite direction (west to east); this was the problem of whether the so-called *diurnal motion* belongs to the Earth or to the rest of the universe. Another question was whether the Sun revolves yearly from west to east around the Earth, or the Earth revolves in the same direction around the Sun; this was the issue of whether the so-called *annual motion* belongs to the Sun or to the Earth. Another aspect of the controversy was whether the center of the universe, or at least the center of the revolutions of the planets, is the Earth or the Sun. And there was also the problem of whether the universe is divided into two very different regions, containing bodies made of different elements, having different properties, and moving and behaving in different ways: the terrestrial or sublunary part where the earth, including water and air, are located; and the celestial, heavenly, or superlunary region, beginning at the Moon and extending beyond to include the Sun, planets, and fixed stars.

The traditional view may be labeled *geostatic,* insofar as it claims the Earth to be motionless; or *geocentric,* insofar as it locates the Earth at the center of the universe; or *Ptolemaic,* insofar as in the second century AD the Greek astronomer Claudius Ptolemy (about AD 100–78) had elaborated it in sufficient detail to make it a workable theoretical system; or *Aristotelian,* insofar as it corresponded to the worldview advanced in the fourth century BC by the Greek philosopher Aristotle, whose ideas in a wide variety of fields had become predominant in the sixteenth century. The other view may be called either *geokinetic,* insofar as it holds the Earth to be in motion; or *heliocentric,* insofar as it places the Sun at the center; or *Copernican,* named after Nicolaus Copernicus, who in the first half of the sixteenth century elaborated its details into a workable theoretical system; or *Pythagorean,* named after the ancient Greek pre-Socratic Pythagoras, who was one of the earliest thinkers (sixth century BC) to advance the idea in a general way. We may thus say that the scientific issue was essentially whether the geostatic or the geokinetic theory is true, or at least whether one or the other is more likely to be true.

The epistemological and methodological issues were several. There was the question of whether physical truth has to be directly observable, or whether any significant phenomenon (e.g., the Earth's motion) can be true even though our senses cannot detect it directly, but can detect only its effects; remember that even today the Earth's motion cannot be seen directly by an observer on Earth. Then there was the question of whether artificial instruments like the telescope have any legitimate role in the search for truth, or whether the proper way to proceed is to use only the natural senses; in fact, the telescope was the first artificial instrument ever used to learn novel scientific or philosophical truths about the world. A third issue of this sort involved the question of the role of the Bible in scientific inquiry, whether its assertions about natural phenomena have any authority, or whether the search for truth about nature ought to be conducted completely independently of the claims contained in the Bible; this was not only a methodological or epistemological issue, but also a theological or hermeneutical one, and it was the paramount issue in the trial, since it was widely believed that the new geokinetic theory contradicted the Bible. Fourth, there was the question of the nature of hypotheses and their role in the search for truth: whether they are merely instruments for mathematical calculation and observational prediction that can be only more or less convenient but neither true nor false, or whether they are assumptions about physical reality that are more or less probable and potentially true or false but not yet known with certainty; here, this problem stemmed from the fact that even the anti-Copernicans admitted that one could explain the motion of the heavenly bodies by means of the hypothesis of the Earth's motion, but they took this as a sign of its instrumental convenience and not of its truth, potential truth, or probable truth. Let us call these four central issues, respectively, the problems of the observability of truth; the legitimacy of artificial instruments; the scientific authority of the Bible; and the role of hypotheses (or the problem of instrumentalism vs. realism).

For the second needed conceptual clarification, one must distinguish between *factual correctness* and *rational correctness*; that is, between being right about the truth of the matter and having the right reasons for believing the truth. Suppose we begin by asking who was right about the scientific issue. It is obvious that Galileo was right and his opponents were wrong, since he preferred the geokinetic to the geostatic view, and today we know for a fact that the Earth does move and is not standing still at the center of the universe. However, it is equally clear that his being right about this fact does not *necessarily* mean that his motivating reasons were correct, since it is conceivable that although he might have chanced to hit upon the truth, his supporting arguments may have been unsatisfactory. Hence, the evaluation of his arguments is a separate issue.

I am not saying that the various proponents of the anti-Galilean accounts are right when they try to show that his arguments left much to be desired, ranging from inconclusive to weak to fallacious to sophistical. In fact, in my opinion, this evaluation is untenable.[4] Rather, I am saying that Galileo's critics have raised a distinct and important issue *about* Galileo's trial—namely, whether, or to what extent, his *reasoning* was correct.

The next distinction that must be appreciated is also easy when stated in general terms but extremely difficult to apply in practice. It is that *essential correctness* must not be equated with either *total correctness* or *perfect conclusiveness*. Applied to our case, this means that even if Galileo's arguments were essentially correct, as I would hold, the possibility must be allowed that the reasoning of his opponents was not worthless, nor irrelevant, nor completely unsound.[5] This point is a consequence of the fact that we are dealing with arguments which logicians would label non-apodictic; that is, they are not completely conclusive, but rather susceptible of degrees of rational correctness. Thus, it is entirely conceivable that there should sometimes be good arguments in support of opposite sides, as well as that the arguments for one side should be better than those for the opposite, without the latter being worthless. I believe this is the case for the trial of Galileo, though it is something the anti-clerical critics do not seem to be able to understand. The proper antidote here is the study of the details of the relevant arguments.

To appreciate the next distinction, let us ask whether Galileo or the Church was right in regard to the epistemological and methodological aspect of the controversy. Since such issues are normally more controversial than scientific ones, this is an area which some like to exploit by trying to argue that the Church's epistemological and philosophical insight was superior to Galileo's. The argument is usually made in the context of a frank and explicit admission that Galileo was unquestionably right on the scientific issue. Thus, these anti-Galilean critics often boast to be displaying even-handedness and balanced judgment by contending that on the one hand Galileo was right from a scientific or factual point of view, but that on the other hand the Church was right from an epistemological or philosophical point of view.

However, such interpretations can be criticized for their exaggeration, one-sidedness, and superficiality in the analysis of the epistemological component of the affair.[6] For example, I have already mentioned that there were at least four epistemological issues in the affair, and I am very doubtful that they can all be reduced to one. Moreover, it cannot be denied that Galileo turned out to be right on at least *some* of the epistemological issues—for example, those pertaining to the legitimacy of artificial instruments and to the Bible lacking scientific authority. On this last point, recall that, as mentioned earlier, it is now more than one hundred years since the Catholic Church officially adopted the Galilean principle that the Bible is an authority only in matters of faith and morals, and not in questions of natural science, with Pope Leo XIII's encyclical on the subject. Furthermore, it seems to me that with the epistemological issues too one can apply the distinction between factual and rational correctness, and thus introduce the question of the rationale underlying the two conflicting positions. That is, we can examine their respective arguments and try to determine which were the better ones, although this is more difficult here than in the case of the scientific arguments.

The main point of this last series of considerations is not to decide the initial question with which they began, but rather to underscore the *multiplicity* of the epistemological issues in the Galileo affair, and to suggest avoiding any one-sided focus on a single one.

Finally, one must bear in mind that this episode was *not* merely an *intellectual* affair. Besides the scientific, astronomical, physical, cosmological, epistemological, methodological, theological, hermeneutical, and philosophical issues, and besides the arguments pro and con, there were legal, political, social, economic, personal, and psychological factors involved. To be sure, it would be a mistake to concentrate on these external issues, or even to devote to them equal attention in comparison with the intellectual issues, for the latter constitute the heart of the episode, and so they must have priority. Nevertheless, it would be equally a mistake to neglect the external, or non-intellectual, factors altogether. To them I now turn.

Non-intellectual Factors

Beginning with personal or psychological factors, it is easy to see that Galileo had a penchant for controversy, was a master of wit and sarcasm, and wrote with unsurpassed eloquence. Interacting with each other and with his scientific and philosophical virtues, these qualities resulted in his making many enemies and getting involved in many other bitter disputes besides the main one that concerns us here. Typically, these disputes involved questions of priority of invention or discovery, and fundamental disagreements about the occurrence and interpretation of various natural phenomena. Major controversies included a successful lawsuit against another scholar for plagiarism in regard to Galileo's invention of a calculating device and its accompanying instructions (1606–7); a dispute with his philosophy colleagues at the University of Padua, where he taught mathematics, about the exact location of the novas that became visible in the heavens in October 1604; a dispute with other philosophers in Florence in 1612 about the explanation of why bodies float in water; a dispute with a German Jesuit astronomer named Christoph Scheiner (1573–1650) about priority in the discovery of sunspots and about their proper interpretation, beginning in 1612 and lasting to the end of their lives; and a dispute with an Italian Jesuit astronomer named Orazio Grassi (1590–1654) about the nature of comets, sparked by the appearance of some of these phenomena in 1618. Given what all this indicates about Galileo's personality, one may wonder how he managed to acquire and keep the many friends and admirers he did.

Moving on to social and economic factors, it should be noted that Galileo was not wealthy. He had to earn his living, first as a university professor, and then under the patronage of the grand duke of Tuscany. During his university career, from 1589 to 1610, his economic condition was always precarious. His university salary was very modest, and this was especially so given that he taught mathematics and thus received only a fraction of the remuneration given to a professor of philosophy. This only compounded other unlucky family circumstances, such as having to provide dowries for his sisters. Galileo was forced to supplement his salary by giving private lessons, by taking on boarders at his house, and by working on and managing a profitable workshop that built various devices, some of his own invention. These financial difficulties eased in the second half of his life when he attained the position of "philosopher and chief mathematician" to the grand duke of Tuscany. However, in this position he was constantly facing a different problem, stemming from the nature of patronage and his relationship to his patron:[7] since the fame and accomplishments of an artist, philosopher, or scientist were meant to reflect on the magnificence of the patron, Galileo was in constant need to prove himself scientifically and philosophically, either by surpassing the original accomplishments that had earned him the position or by giving new evidence for that original worth.

The politics of Galileo's trial has to be understood in the context of the Catholic Counter-Reformation. Martin Luther had started the Protestant Reformation in 1517, and the Catholic Church had convened the Council of Trent in 1545–63. So Galileo's troubles developed and climaxed during a time of violent struggle between Catholics and Protestants. Since he was a Catholic living in a Catholic country, it was also a period when the decisions of that council were being taken seriously and implemented and thus affected him directly. Aside from the question of papal authority, one main issue dividing the two camps was the interpretation of the Bible—both how specific passages were to be interpreted and who was entitled to do the interpreting. The Protestants were inclined toward relatively novel and individualistic or pluralistic interpretations, whereas the Catholics were committed to relatively traditional interpretations by the appropriate authorities.

More specifically, the climax of the trial in 1632–3 took place during the so-called Thirty Years War (1618–48) between Catholics and Protestants.[8] At that particular juncture, Pope Urban VIII, who had earlier been an admirer and supporter of Galileo, was in an especially vulnerable position; thus, not only could he not continue to protect Galileo, but he used Galileo as a scapegoat to reassert, exhibit, and test his authority and power. The problem stemmed from the fact that in 1632 the Catholic side led by the King of Spain and the Bohemian Holy Roman Emperor was disastrously losing the war to the Protestant side led by the King of Sweden, Gustavus Adolphus.

Religion was not the only issue in the war, which was being fought also over dynastic rights and territorial disputes. In fact, ever since his election in 1623, the pope's policy had been motivated primarily by political considerations, such as his wish to limit and balance the power of the Hapsburg dynasty which ruled Spain and the Holy Roman Empire. It had also been motivated by personal interest—that is, cooperation with the French, whose support had been instrumental in his election, and who for nationalistic reasons also opposed the Hapsburg hegemony. In the wake of Gustavus Adolphus's spectacular victories, the Spanish and Imperial ambassadors were accusing Urban of having favored and helped the Protestant cause. They mentioned such things as his failure to send the kind of military and financial support which popes had usually provided on such occasions, and his refusal to declare the war a holy war. There were even suspicions of a more direct understanding with the Protestants. Thus, the pope's own religious credentials were being questioned, and there were rumors of convening a council to depose him.

Then there was what may be called the Tuscan factor, which had at least two political aspects. One was that the Grand Duchy of Tuscany, whose ruler Galileo served, was closely allied with Spain, and so the pope's intransigence with him was in part a way of getting back at Spain. The other was related to the fact that many of the leading protagonists in Galileo's trial were Tuscan: for example, Cardinal Robert Bellarmine, the key figure in the earlier phase of the proceedings, and Pope Urban VIII (of the House of Barberini), the moving force of the later proceedings. Thus the entire episode has some of the flavor of a family squabble.

Finally, another political element involved the internal power struggle within the Church, on the part of various religious orders, primarily the Jesuits and the Dominicans, but to some extent also the Capuchins. In the earlier phase of the trial, in 1615–16, Galileo seems to have been attacked by Dominicans and defended by Jesuits, whereas in the later phase, in 1632–3, it seems that the two religious orders had exchanged roles. It is important to appreciate the significance of such internal dissent: the Church was far from being a monolithic entity.

Just as the political background of the affair involved primarily matters of religious politics, so the legal background involved essentially questions of ecclesiastical, or "canon," law.[9] In Catholic countries, the activities of intellectuals like Galileo were subject to the jurisdiction of the Congregation of the Index and the Congregation of the Holy Office, or Inquisition. In the administration of the Catholic Church, a "congregation" is a committee of cardinals charged with some department of Church business.

The Congregation of the Index was instituted by Pope Pius V in 1571 with the purpose of book censorship. One of its main responsibilities was the compilation of a list of forbidden books (called *Index librorum prohibitorum*). This Congregation was abolished by Pope Benedict XV in 1917, and thereafter book censorship was handled once again by the Congregation of the Holy Office, which had been in charge of the matter before 1571.

The Congregation of the Holy Office, in turn, had been instituted in 1542 by Pope Paul III with the purpose of defending and upholding Catholic faith and morals. One of its specific duties was to take over the suppression of heresies and heretics which had been handled by the Medieval Inquisition; hence, from that time onward, the "Holy Office" and the "Inquisition" became practically synonymous. In 1965, at the Second Vatican Council, its name was officially changed to Congregation for the Doctrine of the Faith.

At the time of Galileo, the Inquisition or Holy Office had a complex bureaucracy; the notion of heresy had been given something of a legal definition; and inquisitorial procedures had been more or less codified.

The Inquisition was the most important and authoritative congregation in the Church. This was reflected partly in the fact that it was the only congregation whose head (called "prefect") was the pope himself. Moreover, its membership consisted of about ten cardinal-inquisitors, which made it the largest congregation. Furthermore, its powers were greater, in the sense that it was not only the supreme judicial tribunal adjudicating particular cases, but also a legislative body whose decisions could enact new laws or change previous ones. Although it usually followed past practice and precedent and various explicitly formulated rules, it was not bound by them.

The Inquisition's bureaucracy was correspondingly numerous and complex. Like most other congregations, it had a secretary, whose position was usually filled by the most senior member of the committee, and whose task was to handle correspondence. However, unlike other congregations, it had a professional staff: the commissary, who played the role of an executive secretary; the assessor, who was the chief legal officer; the prosecutor; and the notary, who was in charge of record-keeping. Each of these had an assistant. Then there were the consultants, who subdivided into two groups: theologians and legal experts. Finally, the Inquisition had offices in all major cities, each headed by a "provincial" inquisitor.

Although the Inquisition dealt with other offenses such as witchcraft, it was primarily interested in two main categories of crimes: formal heresy and suspicion of heresy. The term *suspicion* in this context did not have the modern legal connotation pertaining to allegation and contrasting it to proof. One difference between formal heresy and suspicion of heresy was the seriousness of the offense. For example, a standard Inquisition manual of the time stated that "heretics are those who say, teach, preach, or write things against the Holy Scripture; against the articles of the Holy Faith; ... against the decrees of the Sacred Councils and the determinations made by the Supreme Pontiffs; ... those who reject the Holy Faith."[10] The same manual stated that

> suspects of heresy are those who occasionally utter propositions that offend the listeners ... those who keep, write, read, or give others to read books forbidden in the *Index* and in other particular Decrees; ... those who receive the holy orders even though they have a wife, or who take another wife even though they are already married; ... those who listen, even once, to sermons by heretics.[11]

Another difference between formal heresy and suspicion of heresy was whether the culprit, having confessed the incriminating facts, admitted having an evil intention.[12] Furthermore, within the major category of suspicion of heresy, two main subcategories were distinguished:[13] vehement suspicion of heresy and slight suspicion of heresy; their difference depended on the seriousness of the criminal act. Thus, in effect there were three main types of religious crimes, in descending order of seriousness: formal heresy, vehement suspicion of heresy, and slight suspicion of heresy.

When it came to procedure, there were two ways in which legal proceedings could begin: either by the initiative of an inquisitor, based on publicly available knowledge or publicly expressed opinion; or in response to a complaint filed by some third party, who was required to make a declaration of the purity of his motivation and to give a deposition. Then there were specific rules about the interrogation of defendants and witnesses; how injunctions and decrees were to be worded; how, when, and why interrogation by torture was to be used; and the various kinds of judicial sentences and defendant's abjurations with which to conclude the proceedings.

To summarize, the cultural legacy of Galileo in science, religion, and philosophy can be effectively elaborated by focusing on his trial (its background, proceedings, aftermath, and significance) and by stressing the intellectual developments and issues. However, a balanced approach must be followed, by avoiding the two opposite extremes exemplified by the anti-Galilean and anti-clerical myths, and by not completely overlooking the non-intellectual factors. There is no easy way of doing this, but it is helpful to distinguish scientific from epistemological (or methodological) issues, factual correctness from rational correctness, essential correctness from total correctness, the several epistemological issues from each other, intellectual from external factors, and the several external factors (personal-psychological, social, economic, political, and legal) from each other. However, these distinct aspects are also interrelated, so the point is not to deny their interaction, but to make sure they are not confused or conflated with one another. With these methodological tips in mind, let us move on to look in detail at the background to the trial.

CHAPTER 2

WHEN THE EARTH STOOD STILL

A necessary prerequisite, for understanding the Galileo affair, is acquaintance with its intellectual background. And a key element of this background is provided by the world view that for the preceding two thousand years prevailed in the fields of cosmology, physics, and astronomy. This is the world view which, as we have seen, may be labeled geocentric, geostatic, Ptolemaic, Aristotelian, or Pythagorean.

The geocentric view was not a simple and monolithic entity, but rather was a theory that underwent two thousand years of explicit historical development, comprising about five centuries before and fifteen centuries after the birth of Christ (not to speak of its prehistory). It follows that there are many versions of the theory; for example, Aristotle's and Ptolemy's versions differ not only in emphasis but also in matters of substantive detail. The version expounded below is not a synopsis of any one treatise, but rather a reconstruction of the most widely held beliefs at the beginning of the sixteenth century, in a form useful for the understanding and appreciation of the Galileo affair. My account has been inspired primarily by Galileo's own *Treatise on the Sphere, or Cosmography*, a short elementary textbook of traditional geostatic astronomy which he wrote and used in the early part of his teaching career, but never published.[1]

Cosmology

It is useful to begin with the question of the Earth's *shape*. The geostatic view held that the Earth is a sphere, so that its surface is not flat but round; this is, of course, true. In fact, the evidence and the arguments proving this fact were already known to Aristotle, and they can be found in his writings. Although uneducated people at the time of Aristotle or Galileo may have believed that the Earth is flat, scientists and philosophers had settled the question much earlier. The Copernican controversy had nothing to do with the shape of the Earth, but rather with its behavior, status, and location in the universe.

Similarly, the maritime voyages and geographical discoveries of Christopher Columbus (1441–1506) and others at the end of the fifteenth century and thereafter may have provided additional confirmation of the Earth's spherical shape; but this was only a more direct, experiential proof of the Earth's roundness. On the other hand, those voyages did provide new evidence about the Earth's size, structure, and relative amounts of land and ocean; and this evidence did have an effect on cosmological and astronomical thinking.

The questions of size and shape which became part of that controversy concerned those of the whole universe. In fact, the old view held that the universe was a sphere much larger than the Earth, but of *finite* size, the size being only slightly larger than the orbit of the outermost planet; that is, the distance from the outermost planet to the stars was of the same order of magnitude as the distance between one planet and another. The stars were all at the same distance from the center, being attached to the surface of the so-called *stellar* or *celestial* sphere; this stellar sphere enclosed the whole universe, and outside this sphere there was nothing physical. In other words, the size and shape of the universe were defined in terms of the size and shape of the sphere on which were attached the approximately 6000 fixed stars visible with the naked eye. This contrasts with the modern view that the universe is infinite, space goes on without end, and stars are scattered everywhere in infinite space; so that it does not make much sense to speak of the shape, size, or center of the universe.

Nevertheless, the finite spherical universe was based on the same set of observations which led to the belief that at the center of the celestial sphere was located the motionless Earth. This was the phenomenon of apparent *diurnal motion*: the Earth feels to be at rest; the whole universe appears to move daily around the Earth in a westward direction; this is most obvious for the case of the Sun, whose rising in the east and setting in the west generates the cycle of night and day; but the Moon can also be easily seen to do the same; and thousands of stars visible with the naked eye at night appear to undergo no change in size or brightness, but seem to be at a fixed distance from us; they appear to move in unison, so that their relative positions remain fixed; they appear to move in circles which are larger for stars lying closer to the equator and smaller for those lying closer to the poles; in short, the stars appear to move as if they were attached to a sphere which was rotating daily westward around a motionless Earth at the center. Given the plausible principle that what appears to normal observation corresponds to reality, here was the basic argument in support of the key tenets of the geostatic world view.

In the spherical finite universe, position or location or place had an absolute meaning. The geometrical center of the stellar sphere was a definite and unique place, and so was its surface or circumference; and between the center and the circumference, various layers or spherical shells defined intermediate positions. The part of the universe outside the Earth was called *heaven* in general, and to distinguish one heavenly region from another they spoke of different heavens (in the plural). For example, the stellar sphere could be regarded as the highest heaven, which meant the most distant one from the Earth, and which was also called the *firmament*; whereas the closest heaven was the spherical layer to which the nearest heavenly body (the Moon) was attached, and so the lunar sphere or sphere of the Moon was the *first heaven*. Between the lunar and the stellar spheres, six other particular heavens or heavenly spheres were distinguished: one was for the Sun, and then there was one for each of the other five known planets (Mercury, Venus, Mars, Jupiter, and Saturn). A *heavenly sphere* was not the same as a *heavenly body*: a heavenly sphere was one of the eight nested spherical layers surrounding the central Earth, each of which was the region occupied by a particular heavenly body or group of heavenly bodies, and to each of which these heavenly bodies were respectively attached; whereas a heavenly body was a term referring to the Sun, Moon, a planet, or one of the thousands of fixed stars. The two terms may be confusing because heavenly bodies were considered to be spherical in shape, and so they were spheres in their own right; however, the term "heavenly sphere" referred only to one of the spheres concentric with the center of the universe to which the Sun, Moon, planets, and fixed stars were attached.

The terrestrial region too had its own layered structure. This is related to a three-fold meaning for the term *earth*. In saying earlier that the Earth is a sphere, I was referring to the terrestrial globe consisting of land and oceans. This globe is a sphere, not in the sense of a perfect sphere, but only approximately, because the land is above the water and is full of mountains and valleys. Such an approximation is, of course, very good because the height of even the tallest mountain is insignificant compared to the Earth's radius. However, it was only natural to distinguish water from earth, taking the latter term in the sense of just land, rocks, sand, and minerals; when so understood, "earth" is obviously only a part of the whole globe. Next, it was also natural to count the air or atmosphere surrounding the globe as part of the terrestrial region of the universe; and so by earth one could also mean the whole region of the universe near the terrestrial globe, up to but excluding the Moon and the lunar sphere. In short, the term earth had three increasingly broader meanings: it could refer to just the solid part of the terrestrial globe; or it could refer to the whole globe consisting of both land and oceans; or it could refer to the whole terrestrial region of land plus oceans plus atmosphere.

Terminology aside, the substantive point is that the Earth (namely, the place where mankind lives) is not a body of uniform composition, but has three main parts; it contains a solid part, a liquid part, and a gaseous part. These three parts (namely, earth, water, and air) were labeled *elements* to signify their fundamental importance. In regard to the arrangement of these terrestrial substances, the element earth sinks in water, and so earth must extend to the central inner core of the world and must make up most of what exists below the surface; on the other hand, most of the surface of the globe is covered with water, and the element water mostly surrounds the element earth. This was expressed theoretically by claiming that the natural place of the element earth was a sphere immediately surrounding the center of the universe, and that the natural place of the element water was a spherical layer surrounding the innermost sphere. For the case of the air, simple observation tells us that it surrounds the spheres of the first two elements, and so its natural place was a third sphere surrounding the first two.

There was a fourth terrestrial element, to which the name *fire* was given; but it required a more roundabout explanation. Just as we see earth sink in water, and water fall down (as rain) through air, we see flames shoot upwards through air when something is burning; we also see currents of heat move upwards through air during hot summer days, and smoke generally rise; and we see trapped fire escape upwards in volcanic eruptions. Such observations were taken as evidence that the natural place of fire was a fourth spherical layer above the atmosphere and just below the lunar sphere.

The existence of the element fire was also derived from some considerations about basic physical qualities. There were two fundamental pairs of physical opposites: hot and cold, and humid and dry. The element earth was a combination of cold and dry; the element water was a combination of cold and humid; and the element air was a combination of hot and humid. So there had to be a combination of hot and dry, and that was what constituted the element fire.

In summary, from the point of view of location in the geostatic finite universe, there were twelve natural places, each consisting of a sphere or spherical layer with a common center. The four terrestrial spheres were the natural places of the four terrestrial elements (earth, water, air, and fire). The eight heavenly spheres were the natural places of the heavenly bodies; they ranged from the lunar sphere to the stellar sphere, with six intermediate spheres for the Sun and the five planets. The stellar or celestial sphere enclosed everything else, while the Earth was the center of everything else.

Like position, direction had a definite and absolute meaning in the finite universe. There were three basic directions: toward the center of the universe, which was called *downward*; away from the center of the universe, called *upward*; and around the center of the universe. Thus, one important way of classifying motions was in these cosmological terms: bodies could and did move toward, away from, and around the center of the universe.

Geometrically speaking, motion could be simple or mixed. Simple motion was motion along a simple line. A simple line was defined as a line every part of which is congruent with any other part. Thus, there were supposedly only two such lines, circles and straight lines; and there were two types of simple motion, straight and circular motion. Mixed motion was motion which is neither straight nor circular.

Another way of classifying motions was in terms of the motions characteristic of the various elements, those which the elements undergo spontaneously. Earth and water characteristically moved straight downwards, while air and fire characteristically moved straight upwards. Now, since heavenly spheres and heavenly bodies moved characteristically with circular motion around the center, this meant that they must be composed of a fifth element; the term *aether* or *quintessence* was used to refer to this heavenly element.

Finally, another important classification was in terms of the opposition between natural and violent motions. *Violent motion* was motion caused by some external action; *natural motion* was motion which a body underwent because of its nature, so that the cause was internal. For example, the downward motion of earth and water, the upward motion of air and fire, and the circular motion of heavenly spheres and heavenly bodies were all cases of natural motion; on the other hand, rocks thrown upwards, rain blown sideways by the wind, a cart pulled by a horse, and a ship sailing over the sea were all cases of violent motion.

More fundamentally, motion was the opposite of rest. Rest was the natural state of bodies, and so all motion presupposed a force in some way. Natural motion was essentially the motion of a body toward or within its proper place; only when displaced from its proper place by some force would a terrestrial body engage in natural motion up or down; and only if started by some mover would a heavenly sphere rotate around the center of the universe, thus carrying its planet or stars in circular motion. On the other hand, violent motion was motion which was not toward the body's proper place, and such motion could only happen by the constant operation of a force.

From what has already been said, it is apparent that Earth and heaven were very different; indeed, this radical difference was enshrined in an idea which needs to be made explicit and which deserves a special label. The key term is the *Earth–heaven dichotomy*; but one could equivalently speak of the dichotomy between the earthly or terrestrial or sublunary or elemental region of the universe, on the one hand, and the heavenly or celestial or superlunary or aethereal region on the other.

We have already seen that one difference between the two regions was location, which was absolute in the finite spherical universe: terrestrial bodies occupied the central region of the universe below the Moon, while heavenly bodies occupied the outer region from the lunar to the stellar sphere. Similarly, we have also seen that there was another difference in regard to natural motions: earthly bodies moved naturally straight toward or away from the center of the universe, whereas celestial bodies moved circularly around the same center. We have also seen that the two regions differed in regard to the elements of which bodies were composed. Sublunary bodies were made of earth, water, air, or fire, or a mixture thereof. On the other hand, in the superlunary region things were made of aether, or various concentrations thereof; that is, aether in low concentration made up the heavenly spheres, which were actually invisible; whereas aether in a highly concentrated state generated the Moon, Sun, planets, and stars, which were the heavenly bodies we actually saw.

Now, just as the natural places and the natural motions of the two regions obviously corresponded to each other, the elements in the two regions also corresponded to the natural places and motions. That is, the natural places and the natural motions of terrestrial bodies could be conceived as the essential properties of the terrestrial elements, while the natural places and motions of celestial bodies could be conceived as the essential properties of aether.

Other differences between Earth and heaven could be defined in terms of additional properties of the different elements. For example, whereas superlunary substances had no weight, sublunary bodies obviously did. Or to be more exact, whereas aether was weightless, the sublunary bodies subdivided into two classes: earth and water had weight or *gravity*, and so they were called *heavy bodies*; but air and fire had *levity*, namely the tendency to go up, and so they were called *light bodies*. Moreover, aether was regarded as intrinsically luminous, capable of giving off its own light, while earthly elements were dark. Even fire did not emit an inherent light of its own, but only temporarily produced light when in the process of escaping from lower regions to move to its natural place just below the lunar sphere.

Of the various differences between Earth and heaven, two deserve special attention: natural motion and susceptibility to qualitative change. Natural motion has always been regarded as one of the essential or defining characteristics of a physical body. This is something that seems to have remained unchanged even by the Copernican Revolution; from this point of view, what changed was the natural motion which is attributed to bodies. Since the geocentric theory attributed different natural motions to terrestrial and to celestial bodies, it ought to come as no surprise that it believed in the Earth–heaven dichotomy.

The geostatic universe was *not* a trichotomy, even though there were three visible kinds of natural motions (downward for earth and water, upward for air and fire, and around the center of the universe for aether). One reason was that the downward and upward natural motions were conceived as two minor subspecies of the same fundamental geometrical kind, namely straight or rectilinear motion.

However, this geometrical reason was not the only justification why the essential distinction was the two-fold one between straight and circular natural motions, rather than the three-fold one between upward, downward, and around. There was also the cosmological reason that, unlike circular natural motion, straight natural motion could not be everlasting or perpetual. For once a rock had reached the center of the universe, its nature would make it remain there rather than continue moving past the center, which would constitute upward and thus unnatural motion for the rock. Similarly, once a fiery object had reached the region above the terrestrial atmosphere just below the lunar sphere, it had reached its natural place and had nowhere else to go; for to continue moving would bring it into the first heavenly sphere, which was reserved for the aethereal Moon, and where the element fire could not subsist.

Finally, there was a theoretical reason why upward and downward natural motions could belong to the same fundamental region of the universe, but were essentially different from natural circular motion. The theory in question was the theory of change and contrariety, according to which all change derives from contrariety, and no change can exist where there is no contrariety; contrariety in this context meant oppositions such as those between hot and cold and between dry and humid. Now, up and down, together with the related pair of light and heavy, was another fundamental contrariety. It followed that a region full of bodies, some of which moved naturally downwards and some upwards, was bound to be full of all sorts of qualitative changes; and indeed observation obviously revealed that the terrestrial world is full of births, growth, decay, generation, destruction, weather and climatic changes, and so on. On the other hand, the circular natural motion of the heavenly bodies was thought to have no contrary; consequently, the heavenly region lacked an essential condition for the existence of change.

Add to this that the opposition between hot and cold and between dry and humid belonged only within the four terrestrial elements, and one could claim that the region of aether lacked any of the proper conditions for change. And observation confirmed that too because no physical or organic or chemical changes are easily detected in the heavens, and none were said to have ever been seen. The only essential phenomenon in the heavens was motion, but all heavenly motion was fundamentally regular: it involved the rotation of concentric spheres, which thus remained in place, so that there was not even change of place; what changed was only the relative position of the various bodies attached to the different celestial spheres.

Natural motion and qualitative change, then, provided the basis for the Earth–heaven dichotomy. There were many differences between Earth and heaven, but two interrelated differences were especially important: in the terrestrial world bodies moved naturally with rectilinear motion and underwent all sorts of qualitative changes, whereas in the celestial region things moved naturally with circular motion and were not subject to qualitative change.

To summarize our discussion so far, the Aristotelians and Ptolemaics believed that the Earth was spherical, motionless, and located at the center of the universe; that the universe was finite, bounded at the outer limit by the stellar sphere, and structured into a series of a dozen nested spheres, all inside the stellar sphere and surrounding the central sphere of the solid element earth; that there was a fundamental division in the universe between the earthly and the heavenly regions; and that these regions consisted of bodies with very different properties and behavior, such as different natural places, natural motions, elemental composition, and possibilities for qualitative change. Two things must now be added to this general cosmological picture: the details of the physics of the motion of terrestrial bodies and the astronomical details of the heavenly bodies. Let us begin with the former.

Physics

In the terrestrial region, the *natural state* of bodies was rest. To be more exact, it was rest at the proper place, depending on the elemental composition of the body: at the innermost core for the element earth; just above that for water; above water for air; and above air for fire. This meant that, whereas no cause was sought to explain why a body rested at its proper place, when a body was in motion or at rest outside its proper element, then an explanation was required.

Now, the explanation for why a body was in motion could be that it was going to rest at its proper place; this was the case of natural motion like rocks and rain falling or smoke rising though air. Or the explanation could be that the body was being made to move by an external agent; this was the case of violent motion like a cart pulled by a horse, or a boat sailing over the water, or rain blown by the wind, or weights being lifted from the ground to the top of a building. Both natural and violent motions required a force; the only difference was that in natural motion the motive force was internal to the body, whereas in violent motion the force was external. For example, falling bodies fell because of their inherent tendency to go down, to go to their natural place if they were not already there; the term *gravity* was used to refer to this internal force, and it was measured by the weight of an object. On the other hand, for a sailboat the wind was obviously the external force, and for a cart the horse.

Sometimes "violent motion" was equated with "forced motion," but in such cases it was understood that by "forced motion" one meant motion caused by an *external* as distinct from *internal* force. Since all motion was forced, the term "forced motion" was sometimes regarded as redundant if taken to mean caused motion, and it was found useful only if taken to mean externally caused motion. In other words, the term *force* was ambiguous and could mean either any cause of motion or an external cause of motion; this may cause some confusion, but the context usually clarifies the meaning.

All motion, then, whether natural or violent, was caused by a motive force, whether internal or external. There was, however, another condition which was required by all motion, namely *resistance*. That is, in a sense motion was the overcoming of resistance. This was so in part because all space happened to be filled and there was no vacuum or void, so that whenever a body was moving it could only move through some medium, be it air, water, oil, molasses, sand, or soil. Even the heavenly region, interplanetary and interstellar space, was not devoid of matter; it was filled with (invisible) aether.

Moreover, it was argued that if there were no resistance to overcome, then a motive force (however small) would make a body move instantaneously, namely with infinite speed; and this was an absurdity since it meant that the body would occupy different places at the same time, and indeed many different places at the same time; it followed that there could not be a void, vacuum, or zero resistance. This argument depended on the idea that speed is inversely proportional to resistance, for this idea would provide the justification of why motion without resistance would be instantaneous; that is to say, not only was resistance required for motion to occur, but motion was correspondingly slower with greater resistance and faster with lesser resistance.

This quantitative relationship between speed and resistance was apparently taken seriously for the extreme case of zero resistance and used as just indicated in the above argument. However, the relationship was not taken equally seriously for the other end of the spectrum, for very strong resistance. That is, when the resistance was very strong, rather than saying that a given force would cause some motion, perhaps at very slow speed, it was held that there was a threshold for motion to occur at all; the force had to be sufficient to overcome the resistance in the first place, and if that was the case then the speed was inversely proportional to the resistance. Here, the typical example was that of a single man trying to pull a ship into dry dock by himself; it is clear that he will not be able to move the ship at all, not even 100 times slower than a team of the one hundred men required to accomplish the task.

The relationship between force and speed (when the resistance was constant) was also sometimes expressed quantitatively. The formula was that at constant resistance, the speed is directly proportional to the force. Here the paradigm example was the fall of heavy objects through a fluid like water; heavier objects do sink faster than lighter ones, and do so more or less in proportion to the weight; and weight in this case is the (internal) motive force.

In a modern conceptual framework and using modern terminology, we could combine the two relationships and obtain the following formula: given that the force can overcome the resistance, the body moves at a speed which is directly proportional to the force and inversely proportional to the resistance, that is, *speed* = *constant* × (*force*/*resistance*).

These ideas had great plausibility and were largely in accordance with observation, except for situations like free fall through air and violent projectile motion. For free fall, the Aristotelian theory implied that a lead ball fell much faster than a rock, so that when dropped from the same height the lead would reach the ground much earlier than the rock; moreover, for a given object its speed of fall should not increase with time because it depended only on its fixed weight and the fixed resistance of the air. The problem of projectiles involved the motion of such things as arrows shot from bows, the question being where was the force making them move after the projectiles had left the ejector. The Aristotelians were aware of these problems and tried to solve them, but their solutions were found to be increasingly unsatisfactory. Indeed, it was the discussion of these problems that provided one line of development in the rejection of the old physics and the construction of the new one. However, this was not the only line of development, and, as we shall see later, Galileo's new physics was concerned not only with the problem of falling bodies, but also with the problem of the Earth's motion.

Astronomy

The main astronomical details of the geostatic world view can be visualized in terms of Figure 3.[2]

Imagine a large sphere (NS) surrounding a small one (N′S′) at its center, and let the small sphere represent the Earth and the large one the stellar or celestial sphere. The diurnal motion was conceived as the daily rotation of the large sphere around a line, called the *axis* of diurnal rotation, which went through the north and the south celestial *poles* (N and S); this line also intersected the Earth's center and two points on its surface, the north and the south poles of the Earth (N′ and S′). From an observational viewpoint, the celestial poles were the two points in the heavens that appeared to be motionless (the north celestial pole to observers in the Earth's northern hemisphere, and the south celestial pole to observers in the southern hemisphere); and the circular paths of the fixed stars appeared to be centered at the respective poles. On the surface of the celestial sphere, midway between the poles was a great circle of special importance, called the *celestial equator*; it too had a terrestrial counterpart (the Earth's equator), which could be defined as the intersection of the plane of the celestial equator with the Earth's surface, or as the great circle on the Earth's surface halfway between the north and south terrestrial poles.

One reason for the importance of the celestial poles and equator was that they yielded a fixed frame of reference to define the position of the heavenly

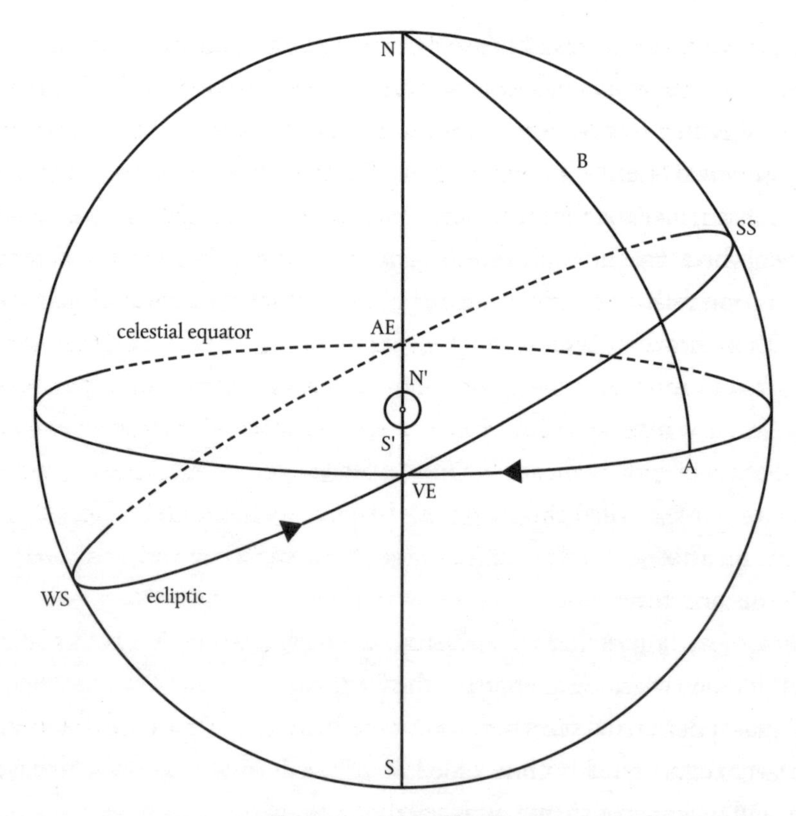

Figure 3. Celestial sphere

bodies, and correspondingly the position of points on the Earth's surface. One could measure the angular position of a star north or south of the celestial equator, which was called *declination* (AB); correspondingly, the angular distance from the terrestrial equator of a point on the Earth's surface is called *latitude*. For each declination or latitude one could conceive a plane parallel to the equator whose intersection with the surface of each sphere generated circles (called *parallels*) that became smaller as one moved toward a pole. On the other hand, the east–west position of a star (B) required first the drawing of a *meridian*, namely, a great circle (partially shown as NBA) through the star and the poles; then one would measure the *ascension*, namely, the angular distance from this meridian to some particular meridian (for example, A to VE); it was analogous for positions on the Earth's surface, except that this east–west angular distance is called *longitude*.

There were two kinds of heavenly bodies, called *fixed stars* and *wandering stars*. A fixed star was a heavenly body that moved daily around the Earth in such a way that its position relative to most other heavenly bodies did not change (it was "fixed"); for example, its declination remained constant, and so did its angular distance from any other fixed star. A wandering star was a heavenly body that not only moved daily around the Earth, but also changed its position relative to other heavenly bodies; that is, the wandering stars were those heavenly bodies which, besides undergoing the diurnal motion, appeared to move in other ways (namely, "wandered" about). There were only seven wandering stars, which were also called planets; indeed, the word *planet* originally meant literally "wandering star." Because wandering stars were often called simply planets, fixed stars were often called simply stars. So although one broad meaning of the word *star* was synonymous with the term *heavenly body*, one narrow meaning of *star* was identical to the term *fixed star*; the term *fixed* was often dropped when the context made it clear that one was indeed referring to fixed stars.

Thousands of fixed stars were visible on a clear night with the naked eye; they were catalogued both in terms of apparent brightness (called *magnitude*) and in terms of shapes or patterns formed by groups of stars close to each other (called *constellations*). The naked eye could be trained to distinguish six magnitudes; stars of the first magnitude were the brightest, and those of the sixth magnitude were the faintest. The brightest star was named Sirius or the Dog Star; it was located near the equator and was part of the constellation of Canis Major. One particular star of the second magnitude was especially important because it was so close to the north celestial pole that, for practical purposes (such as navigation), it could be regarded to be the pole; it was called Polaris or the North Star and was part of the constellation of Ursa Minor.

Both the Sun and Moon were planets because they moved ("wandered") in relation to the fixed stars. Because of their brilliance and their relatively large size, they were called the two *luminaries*. The other known planets were named Mercury, Venus, Mars, Jupiter, and Saturn. We now know that there are other planets circling the Sun in orbits beyond Saturn, but they were unknown not only to the ancients but also to Copernicus and Galileo, and so they played no role in the Copernican Revolution.

The most important point about the planets was that, out of the thousands of heavenly bodies, there were seven that circled the Earth westward once a day like all others, but did not do so in unison with them; these seven bodies also revolved slowly eastward, so that from day to day their position shifted. Whereas a fixed star revolved around the Earth in such a way that after twenty-four hours it returned to the same position (relative to other stars), after twenty-four hours a planet did not quite return to the earlier position but usually had fallen behind somewhat, being located slightly eastward. This can be seen most easily for the case of the Moon by observing its position on succeeding nights at midnight; relative to the fixed stars, it appears to move eastward. In other words, the planets seemed to behave as if their motion were a combination of two circular motions in opposite directions: they circled the motionless Earth westward with the universal diurnal motion, and in addition they simultaneously moved slowly eastward.

The individual planets moved eastward at different rates. The Moon took about a month to return to the same position relative to the fixed stars; the Sun took one year; Mars about two years; and Saturn about twenty-nine years. Thus, the planets moved not only relative to the Earth and the fixed stars, but also relative to each other; each planet had its own distinctive motion, besides the universal diurnal motion. Since the westward diurnal motion was common to all, when one spoke of planetary motions one usually referred to the distinctive individual motions of the planets. Note that, while all the individual planetary motions were usually eastward, this direction was opposite to that of the diurnal rotation, which was westward.

The planetary motion of the Moon, which took about a month, was the most readily observable one since it was connected with the cycle of its phases; a full moon is easily seen and the period from one full moon to the next is an obvious unit of time that can be used as the basis for a calendar. The planetary motion of the Sun was also easy to observe since it is related to the cycle of the seasons of the year; hence, as we have already mentioned, it was called the annual motion.

Everyone can easily observe that, in the course of a year, the rising or setting Sun slowly moves in a north–south direction. Sometimes it rises near due east and sets near due west, which is to say that it is seen on the celestial equator; this happens around March 21, which is the time of the *vernal equinox*; it also occurs around September 23, the time of the *autumnal equinox*. Sometimes it rises and sets about 23.5 degrees north of due east and due west, respectively (namely north of the celestial equator); this happens in the northern hemisphere around June 22, the time of the *summer solstice*. Sometimes it rises and sets about 23.5 degrees south of due east and due west, respectively (south of the celestial equator); this occurs in the northern hemisphere around December 22, the time of the *winter solstice*. One can also observe from a given location on the Earth's surface the elevation above the horizon of the Sun at noon; in the course of a year this elevation changes daily and ranges about 47 degrees, being highest around June 22 and lowest around December 22 (in the northern hemisphere).

This annual northward and southward motion of the Sun indicates that its position relative to the fixed stars changes along a north and south direction since the fixed stars remain at a constant distance from the celestial equator. In other words, the declination of the Sun changes by about 47 degrees during a year, while the declination of a fixed star does not change; so this north–south motion of the Sun is part of its "wandering" among the fixed stars.

Although this apparent northward–southward solar motion was the one most easily observed, it was different from its planetary motion which was eastward. The two were related as follows. The Sun's eastward revolution in its planetary orbit did not take place in the plane of the celestial equator but in a plane inclined to it by 23.5 degrees. The point was that the Sun's motion among the fixed stars was not *exactly* eastward, but *mostly* eastward; its trajectory was slanted north and south. The Sun moved eastward and southward for six months, and eastward and northward for the other six months. The obvious difficulty in observing the Sun's eastward motion among the fixed stars stems from the fact that they cannot be seen when the Sun is visible. What one can do is to observe some star located near the celestial equator and rising in the east soon after the Sun sets in the west; this means that the

Sun and star are diametrically opposed, or about 180 degrees apart. Next, observe the position of the same star just after sunset about a month later; it will be seen to be not just rising, but high in the sky and about 30 degrees west of its previous position; that means that the Sun is now only about 150 degrees away, which is to say that Sun has moved eastward about 30 degrees closer to the star. About six months after the first observation, the star will appear and immediately set in the west just after sunset. Twelve months later, the star will again rise in the east when the Sun sets in the west.

The planetary motion of the Sun may be pictured as in our diagram (Figure 3). Imagine looking at the large sphere from above the north celestial pole, and picture the large sphere rotating clockwise around the motionless small central sphere to represent the westward diurnal rotation of the stellar sphere around the Earth. Next, imagine a great circle on the stellar sphere in a plane cutting the equatorial one at an angle of 23.5 degrees, to represent the Sun's geocentric orbit projected onto the stellar sphere; in accordance with standard terminology, let us use the term *ecliptic* to refer to this geocentric orbit, or the corresponding great circle on the stellar sphere, or the plane on which they both lie. The intersection of the ecliptic and the equator on the stellar sphere defines two special points, called the vernal equinox (VE) and the autumnal equinox (AE); and halfway around the ecliptic between the equinoxes are two other special points, the summer solstice (SS) at the northern end, and the winter solstice (WS) at the southern end; these four points thus divide the ecliptic circle into four equal quadrants. Now, imagine the Sun moving counterclockwise around the ecliptic at a rate that makes it traverse the whole circumference in one year; then the Sun will be at VE around March 21, at SS around June 21, at AE around September 21, and at WS around December 21.

Let us now combine the clockwise rotation of the whole stellar sphere with the counterclockwise revolution of the Sun along the ecliptic. The result is that the Sun in reality moved in a helical path which in one year looped clockwise around the Earth about 365 times (days of the year), but which in any one day corresponded almost but not quite to one of the parallels on the stellar sphere. I say "almost but not quite" first because the parallel circle was not completely traversed by the Sun, but fell short by about one degree (1/360 of a circle, which approximately equals 1/365 of a year); and second because the end of the daily path rises northward or drops southward relative to the beginning of the same daily path by 1/4 of a degree on the average (namely 23.5 degrees every 3 months, or every 90 days).

The ecliptic was important not only because it represented the yearly eastward path of the Sun among the stars, but also because it was used to define a frame of reference, distinct from the equatorial one mentioned earlier. For example, one could draw a line perpendicular to the center of the ecliptic (called the axis of the ecliptic); one could then speak of the poles of the ecliptic as the points where its axis intersected the celestial sphere; one could define the position of a star in terms of its angular distance from the ecliptic toward one of its poles; and one could also plot the position of a body in terms of east–west position along the ecliptic.

This ecliptic frame of reference was especially important for the other six planets because they are never seen to wander much away from the ecliptic; that is, planets are always observed to be somewhere inside a narrow belt extending 8 degrees above and below the ecliptic. This was the result of the fact that the individual circular paths of the planets took place in planes which, while not identical with the ecliptic, intersected it at small angles no larger than 8 degrees. This narrow belt on the stellar sphere along which the planets revolved was called the *zodiac*. It was subdivided into 12 equal parts of 30 degrees each, and each part happened to be the location of a group of stars that seemed to be arranged into a distinct pattern. These twelve patterns were the constellations of the zodiac and were named Aquarius, Pisces, Aries, Taurus, Gemini, Cancer, Leo, Virgo, Libra, Scorpio, Sagittarius, and Capricorn. The Sun, Moon, and other planets were at all times found somewhere in one of these constellations, and they moved from one constellation to the next in the order just listed. This order corresponded to what we have called an eastward direction (from the viewpoint of terrestrial observation), or counterclockwise (in connection with the pictorial diagram just described); the key point, however, was that the order of the signs of the zodiac was a direction of motion opposite to that of the diurnal rotation.

When projected onto the stellar sphere, the eastward motion of the planets could be described in terms of great circles on the surface of that sphere, all of which were within the zodiac and intersected one another at small angles. But the planets were not believed to be attached to the stellar sphere like the fixed stars; unlike the fixed stars, the planets were not regarded to be equidistant from the Earth. The fact that the planets appeared to move relative to the fixed stars, and that this motion took place at different rates for different planets, implied that each planet was attached to its own sphere which rotated eastward at its own rate, while being carried westward daily by the diurnal rotation of the stellar sphere.

Except for the Moon (whose distance was relatively ascertainable because of eclipses), there was no direct way to measure the sizes of the various planetary spheres or orbits, but the relative determination was done on the basis of the length of time required for a given planet to complete one circular journey among the stars. The principle used was that the bigger ones of these nested planetary spheres rotated at slower rates, and the smaller ones at faster rates; that is, the bigger the orbit, the slower the period of revolution. This principle was combined with the observation that the periods of revolution ranged from one month for the Moon to one year for the Sun and twenty-nine years for Saturn. The result was that in order of increasing distance from the Earth, the planets were most commonly arranged as follows: Moon, Mercury, Venus, Sun, Mars, Jupiter, and Saturn. Thus, as mentioned earlier, between the stellar sphere and the Earth, there were seven other nested spheres, whose rotation carried the corresponding planets in their own individual eastward orbits, while they were all being carried in a westward daily whirl by the diurnal rotation of the celestial sphere.

Besides the fundamentals just sketched, the geostatic world view also contained some more specific and technical points, but the discussion of these is best postponed until later, as they become relevant in the context of understanding various parts of the geokinetic, heliocentric, and Copernican world view. For now, let me end this sketch by stressing a very important point: the geostatic and geocentric system of Aristotle, as elaborated more technically by Ptolemy, yielded plausible explanations and useful predictions of celestial phenomena; in short, it worked. For about two thousand years, no one was able to come up with anything better. All this changed with Copernicus, to whom we turn in the next chapter.

CHAPTER 3

THE COPERNICAN CONTROVERSY (1543–1609)

Copernicus's Innovation

In 1543, the Polish astronomer Nicolaus Copernicus published an epoch-making book, *On the Revolutions of the Heavenly Spheres*. It so happened that he died the same year; in fact, he received the first copies of the printed book on his deathbed. (Some people are keen to point out that Copernicus both published and perished!)

Be that as it may, in his book Copernicus elaborated the idea that, rather than standing still at the center of the universe, with all the heavenly bodies revolving around it, the Earth moves by rotating on its own axis daily, and by revolving around the Sun once a year. Copernicus was trying to replace the traditional geostatic and geocentric theory with a geokinetic and heliocentric theory.

As in the geostatic view, for Copernicus the Earth was spherical and the universe was finite and spherical; the fixed stars were attached to the celestial sphere and equidistant from the center. However, the celestial sphere was motionless and did not revolve around the Earth with westward diurnal rotation; instead, the diurnal rotation belonged to the Earth, and its direction was eastward, resulting in the observational appearance of the whole universe rotating westward. Copernicus also gave the Earth a second motion, an orbital revolution around the Sun with a period of one year, and also in an eastward direction. That is, the annual motion was shifted from the Sun to the Earth (with the direction remaining unchanged), thus making the Earth a planet, rather than the Sun. This terrestrial orbital revolution meant that the Earth was located off center, the center of the universe being instead the Sun. Moreover, the other five planets continued to be planets, but their orbits were centered on the Sun rather than on the Earth. Around the Sun there thus revolved six planets in the order: Mercury, Venus, Earth, Mars, Jupiter, and Saturn. The Moon remained a body that circles the Earth eastward once a month.

What Copernicus accomplished was to update an idea which had been advanced in various forms by the Pythagoreans, by Aristarchus of Samos (about 310–250 BC), and by other astronomers in ancient Greece, but which had been almost universally rejected. In a sense, Copernicus's accomplishment was to give a *new* argument in support of this *old* idea that had been considered and rejected earlier. His theory was not primarily based on new observational evidence, but was essentially a novel and detailed reinterpretation of available data. He demonstrated in quantitative detail that the *known* facts about the motions of the heavenly bodies (especially the planets) could be explained *more simply* and *more coherently* if the Sun rather than the Earth is assumed to be at the center, and the Earth is taken to be the third planet circling the Sun yearly and spinning daily on its own axis.

For example, there are thousands fewer moving parts in the geokinetic system than in the geostatic one; for the apparent daily motion of all heavenly bodies around the Earth is explained by the Earth's axial rotation, so there is only one thing moving daily (the Earth), rather than thousands of stars. Thus, insofar as simplicity depends on the number of moving parts, the geokinetic arrangement is simpler than the geostatic system.

A similar point can be made in regard to the number of directions of motion. Fewer are needed in the Copernican than in the Ptolemaic system. In the geostatic system there are *two* opposite directions, but in the geokinetic system all bodies rotate or revolve in the same direction. That is, in the geostatic system, while all the heavenly bodies revolved around the Earth with the diurnal motion from *east to west*, the seven planets (Moon, Mercury, Venus, Sun, Mars, Jupiter, and Saturn) simultaneously also revolved around it from *west to east*, each in a different period of time. However, in the geokinetic system, only one direction of motion is needed. For, if the apparent diurnal motion from east to west is explained by attributing to the Earth an axial rotation, then the direction of the latter has to be reversed (west to east). On the other hand, if the apparent revolution of the Sun from west to east is explained by attributing to the Earth an orbital revolution around the Sun, then *the same direction* has to be retained; this is easy to understand by visualizing the situation, as follows. In Figure 4, S represents the Sun; the small circle is the Earth's heliocentric orbit, which the Earth traverses in a counterclockwise (or west-to-east) direction; the large circle represents the stellar sphere, which is regarded as motionless. The key issue is that, as the Earth actually moves through points E1, E2, E3, E4, etc. in its orbit around the Sun, from the Earth the Sun appears to move through a corresponding set of points (S1, S2, S3, S4, etc.) on the stellar sphere; but the direction of motion is the same, counterclockwise (or west-to-east).

The explanatory coherence of the Copernican theory—its capacity to explain many phenomena without having to add artificial and *ad hoc* assumptions—derived from its ability to explain the various known facts about the motions and orbits of the planets by means of basic principles

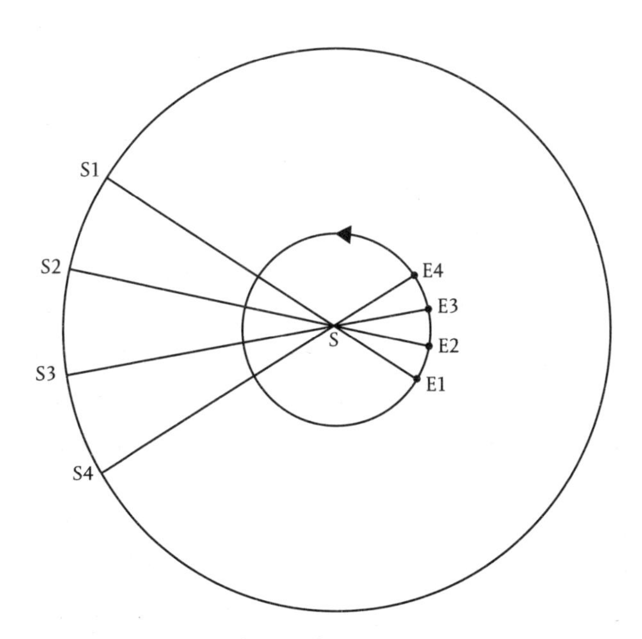

Figure 4. Direction of annual motion in Copernican system

concerning the motion of the Earth. By contrast, in the geostatic system, the thesis of a motionless central Earth had to be combined with a whole series of unrelated assumptions in order to explain what is observed to happen.

The best example of explanatory coherence is the Copernican explanation of retrograde planetary motion and of planetary variation in brightness. Careful observation revealed that no planet moved at a uniform rate in its orbit, but that its speed appeared to vary. Moreover, although the Sun and Moon always moved eastward in their apparent planetary revolutions, the other five planets were periodically seen to slow down, stop, reverse course, and briefly move westward relative to the fixed stars; this reversed movement was called *retrogression* or *retrograde* planetary motion. Finally, during retrogression planets appeared brighter, as if they were nearer the Earth than at other times.

These observations meant that a planet could not be simply attached to a rotating heavenly sphere, for in that case neither the distance nor the direction of revolution nor the speed should change. In the geostatic system, the device most commonly used to explain retrograde motion and variation in brightness and speed was a mechanism consisting of *deferents* and *epicycles* (Figure 5).

A deferent was defined as a geocentric circle whose circumference (ABCD) rotated around the Earth (E). An epicycle was defined as a circle (FGHI) whose center (A) lay on the circumference of the deferent, and whose circumference rotated in the same direction as the deferent. The planet was located on the circumference of the epicycle. Thus, when the rotation of the epicycle carried the planet on the far side (F) of the epicycle from the Earth, its distance was the sum of the radii of the deferent and the epicycle; whereas when the epicyclic rotation carried the planet on the near side (H) of the epicycle from the Earth, its distance was the difference between the two radii. Thus, in its geocentric revolution, the distance of the planet from the Earth changed by an amount equal to the diameter of the epicycle. This difference accounted for the variation in brightness.

Moreover, the planet's motion was the result of its motion along the epicycle and the motion of the center of the epicycle along the deferent. Thus,

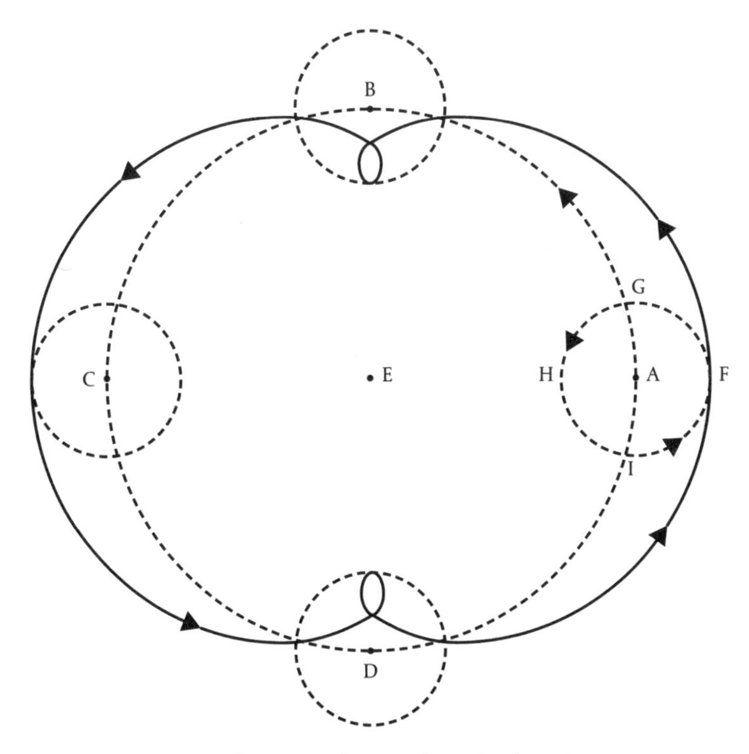

Figure 5. Deferents and epicycles in Ptolemaic system

when the planet was on the far side (F), its speed was the sum of the deferent speed and the epicycle speed; then its speed was faster than its average speed. But when the planet was on the near side (H) of the deferent, its speed was the difference between the two; in this case, if the epicycle speed was greater than that of the deferent, the planet appeared to move backwards (clockwise or westward). Retrograde planetary motion then resulted.

For each planet, the relative sizes of deferent and epicycle and their relative rates of rotation could be adjusted so that their combination yielded mathematically the observed details about retrogression and changes in brightness and speed. For example, if the planet was observed to retrogress twice while revolving through its complete orbit once, then the epicycle was assumed to rotate twice as fast as the deferent; this yielded a path which in reality was looped; but from the Earth (E) the loop was not seen, and instead the planet would appear brighter and retrogressing near B and D.

The framework of deferents and epicycles was a very powerful instrument for the analysis of planetary motion. There was much more that an astronomer could do besides adjusting the relative sizes and speeds of a deferent and its epicycle. For example, one could add a second epicycle on the first epicycle; one could make the center of the deferent different from the center of the Earth, in which case the deferent was called an *eccentric*; and one could even make the center of the deferent move in some way, perhaps in a small circle around the Earth's center. For many centuries before Copernicus, such calculations, adjustments, and refinements involving deferents, epicycles, and eccentrics constituted the primary theoretical and mathematical task of planetary astronomy. This enhanced the power of the theory, but it also rendered the whole system more complicated and *ad hoc*. Moreover, the physical reality of deferents, epicycles, and eccentrics became increasingly unclear.

In the Copernican system, the observed phenomena are explained without the need for Ptolemaic *ad hoc* postulation and construction of epicycles; instead, they are direct consequences of the relative motion between the Earth and the other planets as they all rotate around the Sun. When the Earth and another planet reach points in their orbits which are on the same side of the Sun and are thus at the minimum distance from one another, their different orbital speeds make the other planet seem to move backward (westward).

If the other planet is a superior one, i.e., one with an orbit larger than the Earth's, this retrograde motion is seen when the planet's apparent position on the celestial sphere is opposite to that of the Sun; the Earth's faster orbital motion toward the east leaves behind the other planet, which thus appears to move toward the west relative to the fixed stars. For example, in Figure 6a, while the Earth moves through points E1 to E11 along the smaller orbit, the superior planet moves along its bigger orbit through a corresponding set of points (P1 to P11). The latter make up a smaller portion of its orbit due to its slower orbital speed. Thus, the apparent position of the superior planet against the background of the fixed stars changes in the order S1 to S11; and in this sequence, the part S4, S5, S6, S7, and S8 is retrograde.

(a) Retrogression of superior planet in Copernican system

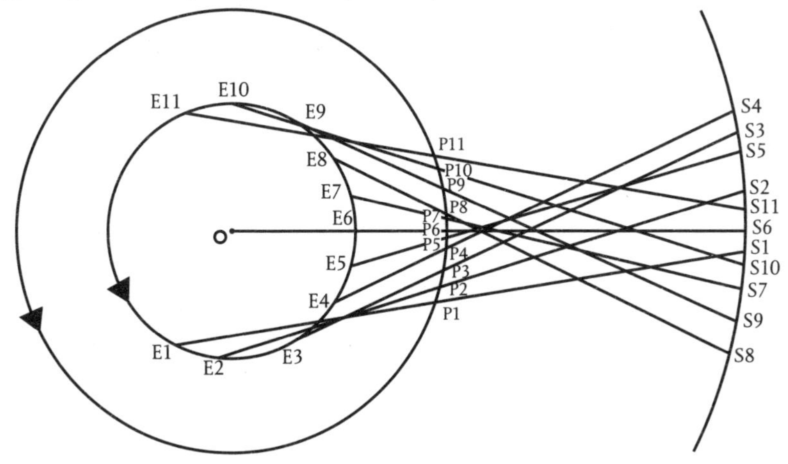

(b) Retrogression of inferior planet in Copernican system

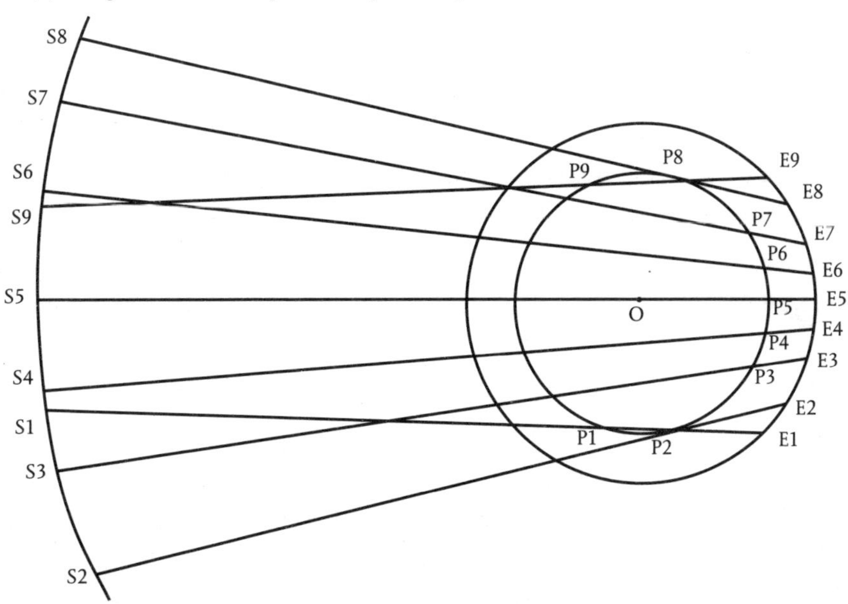

Figure 6. Copernican explanation of retrograde planetary motion

If the other planet is an inferior one, with an orbit smaller that the Earth's, then at certain times the Earth's slower orbital motion enables the other planet to overtake the Earth relative to the fixed stars, thus generating the appearance that the planet is moving in a direction opposite to its usual one. That is, as Figure 6b shows, while the Earth moves counterclockwise through points E1 to E9 along the bigger orbit, the inferior planet moves also counterclockwise along the smaller orbit through a corresponding set of points (P1 to P9); then the apparent position of the inferior planet along the stellar sphere changes in the order S1 to S9, and the part S2, S3, S4, S5, S6, S7, and S8 is clockwise, or retrograde.

Despite these advantages of the geokinetic theory from the points of view of simplicity and explanatory coherence, as a *proof* of the Earth's motion, Copernicus's argument was far from conclusive. Notice first that his argument is a hypothetical one. It is based on the claim that *if* the Earth were in motion, *then* the observed phenomena would result; but from this it does not follow that the Earth is in motion; all we would be entitled to infer is that the Earth's motion offers an explanation of observed facts, and one that was simpler and more coherent than the geostatic explanation. This does provide *two reasons* for preferring the geokinetic idea, but they are not decisive reasons. They would be decisive only in the absence of reasons for rejecting the idea. One has to look at counter-arguments—and there were plenty of them.

The Anti-Copernican Arguments

The arguments against the Earth's motion can be classified into various groups, depending on the branch of learning or type of principle from which they stemmed. In fact, these objections reflected the various traditional beliefs which contradicted or seemed to contradict the Copernican system. The objections were epistemological, philosophical, theological, religious, physical, mechanical, astronomical, and empirical.

Two sets of issues can also be distinguished, each applying to both the ancient and the Copernican view. The ancient view, recall, contained two main parts: the geostatic thesis that the Earth is motionless, and the geocentric thesis that the Earth is located at the center of the universe. These are independent of each other because there is no contradiction in holding that the Earth is devoid of any motion but is located slightly off the center of the universe; this is what some ancient thinkers speculated in order to account for some of the specific details of the apparent motions of the heavenly bodies. Conversely, there is no incompatibility in holding that the Earth is located at the center of the universe but moves by performing a simple daily axial rotation; in fact, this is precisely the kind of compromise position which other thinkers conceived of. Similarly, in regard to the Copernican view, the Earth's daily axial rotation and its annual orbital revolution are distinct, in the sense that one could admit terrestrial rotation but deny orbital revolution; this could be done by letting the Earth rotate at the center of the universe. On the other hand, if one lets the Earth revolve annually around the Sun then it would be absurd to deny terrestrial rotation because this would mean that the apparent daily motion of the heavens was actual, and thus we would have a situation where the Earth was going around the Sun once a year while the whole universe was revolving around the Earth once a day.

Let's begin with the empirical objections, to underscore the fact that the opposition to Copernicanism was neither all mindless nor simply religious. However, to set the stage for the empirical details, it is best to begin with an argument which is empirical in the sense of involving observation and sense experience, but which does so in such a way that what we really have is an epistemological objection.

The argument was aptly called the *objection from the deception of the senses.* To understand the deception involved, note that Copernicus did not claim that he could feel, see, or otherwise perceive the Earth's motion by means of the senses. Like everyone else, Copernicus's senses told him that the Earth is at rest. Therefore, if his theory were true, then the human senses would not be reporting the truth, or would be "lying" to us. But it was regarded as absurd that the senses should deceive us about such a basic phenomenon as the state of rest or motion of the terrestrial globe on which we live. In other words, the geokinetic theory seemed to be in flat contradiction with direct sense experience and to violate a fundamental epistemological principle: that under normal conditions the senses are reliable and provide us with one of the best instruments to learn the truth about reality.

One could begin trying to answer this difficulty by saying that deceptions of the senses are neither unknown nor uncommon; this is shown, for example, by a straight stick half immersed in water that appears bent, or by the shore appearing to move away from a ship to an observer standing on the ship and looking at the shore. Still, the difference is that these perceptual illusions involve relatively minor and secondary experiences, whereas to live all one's life on a moving globe without noticing it would be a gigantic and radical deception. Moreover, it was added, the former illusions are corrigible, since we have other ways of discovering what really happens, whereas there is no way of correcting the perception of the Earth being at rest. This difficulty may be labeled an epistemological objection because the real issue is whether the Earth's motion ought to be (directly) observable, and whether the human senses ought to be capable of directly revealing the fundamental features of physical reality.

This general empirical objection is in a sense the reverse side of the coin of the fundamental advantage of the geostatic system. (The same applies, of course, to all the other anti-geokinetic objections, which may thus be easily turned into pro-geostatic arguments.) The most basic and important argument in favor of the geostatic view was taken from direct observation, which testifies to the correctness of the geostatic thesis: our visual experience reveals that the heavenly bodies move around the Earth every day, a point which is most easily observable for the Sun, whose rising in the east and setting in the west generates the cycle of night and day; further, according to our kinesthetic sense the Earth is felt to be at rest; the argument here was simply that the Earth must be standing still because our sense experience shows this.

The other empirical objections to Copernicanism were more specific, and were based primarily on effects in the heavens which ought to be observed in a Copernican universe, but which in fact were not. These specific empirical difficulties may therefore be also called the astronomical objections.

We can start with the *objection from the Earth–heaven dichotomy*. It argued that, if Copernicus were right, then the Earth would share many physical properties with the other heavenly bodies, especially the planets, since the Earth would itself be a planet, the third one circling the Sun. However, as we have seen (Chapter 2), it was widely believed that whereas the heavenly bodies were weightless, luminous, changeless, and made of the element aether, terrestrial bodies were dark, subject to constant changes, and made of parts that had weight (earth and water) or levity (air and fire). Now, before the invention of the telescope this belief had considerable empirical support.

The issue of *the phases of the planet Venus* was the basis of another objection. If the Copernican system were correct, then this planet should exhibit phases similar to those of the Moon but with a different period; yet none were visible (before the telescope). The reason why Venus would have to show such phases stems from the fact that, in the Copernican system, it is the second planet circling the Sun, the Earth is the third, and these two planets have different periods of revolution. Therefore, the relative positions of the Sun, Venus, and the Earth would be changing periodically, and corresponding changes would occur in the amount of Venus's surface visible from the Earth: when Venus is on the far side of the Sun from the Earth, its entire hemisphere lit by the Sun is visible from the Earth, and the planet should appear as a disk full of light (like a full moon, though much smaller); when Venus comes between the Sun and the Earth, none of its hemisphere lit by the Sun is visible from the Earth, and the planet would be invisible (as in the case of a new moon); and at intermediate locations, when the three bodies are so positioned that the line connecting them forms a noticeable angle, then different amounts would be visible, giving Venus an appearance ranging from nearly fully lit, to half lit, to a crescent shape.

This objection assumes that Venus is dark and opaque, and not intrinsically luminous. However, this is a reasonable assumption for Copernicanism, according to which Venus shares these optical properties with the Earth, the Moon, and the other planets. By contrast, in the Ptolemaic system, Venus is intrinsically luminous, like all other heavenly bodies; thus, its appearance would never show phases, despite its constant location between the Earth and the Sun. It follows that the objection from the phases of Venus is indeed a difficulty for Copernicanism, but not for the Ptolemaic system.

The *apparent brightness and size of the planet Mars* involved another problematic issue. In the Copernican system, this planet revolves in the next outer orbit (M_1–M_2 in Figure 7) after the Earth's orbit (E_1–E_2). Since they also revolve at different rates, they are relatively close to each other when their orbital revolutions align both on the same side of the Sun (E_1 and M_1, or E_2 and M_2), and relatively far when they are on opposite sides of the Sun (E_1 and M_2, or E_2 and M_1). This variation in distance between the Earth and Mars is considerable; according to some estimates, it was supposed to be eightfold. This change in distance would cause a corresponding variation in the apparent size of Mars when seen from the Earth, and an even greater change in brightness, since the intensity of light varies as the square of the distance. Now, the difficulty was that, although Mars did indeed exhibit a noticeable change in brightness with periodic regularity, this change was not nearly as much as it should be; further, there was practically no variation in apparent size (before the telescope).

Again, in the Ptolemaic system the distance between the motionless central Earth and Mars also changed, due to its epicycles; and so its apparent

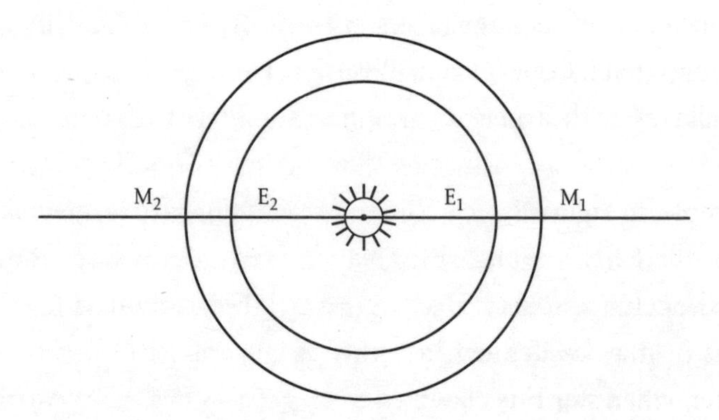

Figure 7. Copernican variations in appearance of Mars

size and brightness should also undergo some changes. However, the appearance of Mars did not present a serious difficulty for the geostatic view, as it does for Copernicanism, because in the Ptolemaic system, the relevant quantities (distance, epicycle, and so on) could be adjusted to correspond to the actual observations, whereas in the Copernican system the variation could be derived from other elements of the system, because of its greater coherence.

A final empirical astronomical argument was based on the fact that observation revealed no change in the *apparent position of the fixed stars*; this is commonly known as the objection from stellar *parallax*, a term that denotes a change in the apparent position of an observed object due to a change in the location of the observer. At its simplest level, the apparent position of a star may be thought of as its location on the celestial sphere, which in a sense is its position relative to all the other stars (also located on that sphere); or, from the viewpoint of the Copernican system, it may be conceived as measured by the angular position of the star above the plane of the Earth's orbit (the so-called plane of the ecliptic). Now, if the Earth were revolving around the Sun, then in the course of a year its position in space would change by a considerable amount, defined by the size of the Earth's orbit; therefore, a terrestrial observer looking at the same star at six-month intervals would be observing it from different positions, the difference being a distance equal to the diameter of the Earth's orbit; consequently, the same star should appear as having shifted its position either on the celestial sphere or in terms of its angular distance above the plane of the Earth's orbit. It follows that if Copernicanism were correct, we should be able to see stellar parallaxes with a periodic regularity of one year. Yet none were observed.

For example, in Figure 8, let ANBO represent the Earth's orbit; line AB a diameter of the Earth's orbit; ABC a line in the plane of that orbit; CEH a portion of the celestial sphere; H a fixed star whose position, when observed from point B, may be defined in terms of the angle HBC. However, six months later, when star H is observed from point A in the Earth's orbit, the star's position may be defined in terms of angle HAC; and this angle (HAC)

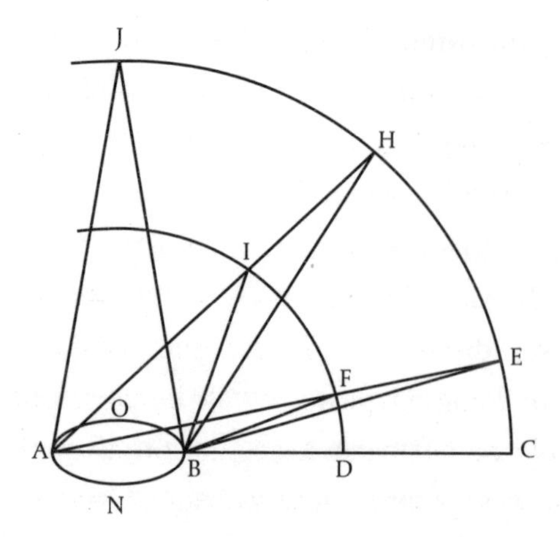

Figure 8. Annual stellar parallax in Copernican system

is smaller than the previous one (HBC). That is, when observed at six-month intervals, the same star H would appear to shift its position, appearing sometimes higher and sometimes lower above the plane of the ecliptic.

As we shall see later, the first three of these empirical astronomical objections were not answered until Galileo's telescopic discoveries, and stellar parallax was not detected until much later, in 1838, by German astronomer Friedrich Bessel (1784–1846). In fact, the magnitude of parallax varies inversely as the distance of the observed object; and the stars are so far away that their parallax is exceedingly small; so for about two centuries telescopes were not sufficiently powerful to make the fine discriminations required. One may then begin to sympathize with Copernicus's contemporaries, including Galileo, who initially found his idea very hard to accept.

Besides, there were many other reasons for their opposition to Copernicanism. The next group of objections may be labeled *mechanical* or *physical*, in the sense that they are based directly or indirectly on a number of principles of the branch of physics which today we call mechanics, which studies how bodies move. We will look first at four objections which hinge indirectly (though crucially) on the laws of motion, and later at two others where the appeal to such physical principles is direct and explicit.

The *objection from vertical fall* began with the fact that bodies fall vertically. This is something that everyone can easily observe by looking at rainfall when there is no disturbing wind; or by throwing a small rock directly upwards, and noticing that it falls back to the place from which it was thrown; or by dropping a rock from the top of a building or tower and observing that it moves perpendicularly downwards, landing directly below. It was argued that this could not happen if the Earth were rotating; for, while the body was falling through the air, the ground below would move a considerable distance to the east (due to the Earth's axial rotation), and although the building and person would be carried along, the unattached falling body would be left behind; so that on a rotating Earth the body would land to the west of where it was dropped, and it would appear to be falling along a westward slanted path. Since this is not seen, but rather bodies are observed to fall vertically, it was concluded that the Earth does not rotate.

An analogous argument was advanced by the *objection from east–west gunshots.* The relevant observation here was that, when ejected with equal force, projectiles range equal distances to the east and to the west. This can be most easily observed by throwing a rock with the same exertion in both opposite directions in turn, and measuring the two distances; one could also use bow and arrow, so as to have a slightly better measure of the propulsive force; or one could use a gun, and shoot it first to the east and then to the west with the same amount of charge. Now, the argument claimed that on a rotating Earth such projectiles should instead range further toward the west than toward the east, because in its westward flight the projectile would be moving against the Earth's rotation, which would carry the place of ejection and the ejector some extra distance to the east; whereas, in its eastward flight, the projectile would be traveling in the same direction as the ejector, due to the latter being carried eastward by the Earth's rotation; therefore, on a rotating Earth the westward projectiles would range further by a distance equal to the amount of the Earth's motion, while the eastward ones would fall short by the same amount. Again, since observation reveals that this is not so, it supposedly follows that the Earth does not rotate.

Of course, today these arguments can be refuted. However, their refutation requires knowledge of at least two fundamental principles of mechanics, to which Galileo himself contributed. One is the law of conservation of momentum, or more simply the principle of conservation of motion, according to which the motion acquired by a body is conserved unless an external force interferes with it. The other is the principle of superposition, which specifies how motions in different directions are to be combined with each other to yield a resultant motion. The point that needs to be stressed is that, since the phenomena to which these two objections appealed to are indeed true, the issues they raised were about how bodies would or could move on a rotating Earth, and the resolution of these issues depended on the possession of more accurate mechanical principles. The next objection raised these same issues, but also the question of what the facts of the case really were; however, to establish these facts was not so easy as it might seem.

The *objection from the ship's mast experiment* referred to an experiment to be made on a ship, and it then drew an analogy between the Earth and the ship. The experiment consisted of dropping a rock from the top of a ship's mast, both when the ship is motionless and when it is advancing forward, and then checking the place where the rock hits the deck. It was asserted that the experiment yielded different results in the two cases: that when the ship was standing still the rock fell to the foot of the mast, but that when the ship was moving forward the rock hit the deck some distance toward the back. Then the moving ship was compared to a portion of land on a rotating Earth, and a tower on the Earth was regarded as the analogue of the ship's mast. From this it was inferred that, if the Earth were rotating (eastward), then a rock dropped from a tower would land to the west of the foot of the tower, just as on a ship moving forward it falls toward the back; however, since the rock can be observed to land at the foot of the tower, they concluded that the Earth must be standing still.

This objection partly involves the empirical issue of exactly what happens when the experiment is made on a moving ship. If the experiment is properly made, the result will be that the rock still falls at the foot of the mast. However, it is easy to get the wrong result due to extraneous causes, such as wind and the rocking motion which the boat is likely to have in addition to its forward motion. Therefore, it is not surprising that there were common reports of the experiment having been made and having yielded anti-Copernican results. Nor is it surprising that, as we shall see later, when Galileo tried to refute the objection, although he disputed the results of the actual experiment, claiming to have performed it, he emphasized a more theoretical answer in terms of the principles of conservation and superposition of motion. These principles are needed to determine what will happen to the horizontal motion the rock had before it was dropped from the mast of the moving ship, and how it is to be combined with the new vertical motion of fall it acquires.

The last one of the indirectly mechanical objections to be discussed here is *the objection from the extruding power of whirling*, or, as we might say today, the *centrifugal-force argument*. The basis of this objection was the fact that in a rotating system, or in motion along a curve, bodies have a tendency to move away from the center of rotation or of the curve. For example, if one is in a vehicle traveling at a high rate of speed, whenever the vehicle makes a turn one experiences a force pushing one away from the center of the curve defined by the turn: if the vehicle turns right, one experiences a push to the left, and vice versa. Or you could tie a small pail of water at the end of a string and whirl the pail in a vertical circle; now suppose a small hole is made in the bottom of the pail; as the pail is whirled you would see water rushing out of the hole always in a direction away from your hand. Then the argument called attention to the fact that, if the Earth rotates, bodies on its surface are traveling in circles around its axis at different speeds depending on the latitude, the greatest speed being about 1000 miles per hour at the equator. This sounds like a very high rate of speed, which would generate such a strong extruding power that all bodies would fly off the Earth's surface, and the Earth itself might disintegrate. Since this obviously does not happen, it was concluded that the Earth must not be rotating.

This objection raised issues whose resolution involved the correct laws of centrifugal force. At the time, however, these laws were not known, and so this objection was a very strong one. Next, we come to the objections according to which the conflict with physical principles was so explicit that the Earth's motion seemed a straightforward physical impossibility.

One of these objections was the *natural-motion argument*. It claimed that the Earth's motion (whether of axial rotation or orbital revolution) is physically impossible because the natural motion of earthly bodies (rocks and water) is to move in a straight line toward the center of the universe. The context of this argument was the science of physics which I elaborated earlier (Chapter 2) as part of the geostatic world view: it contrasted natural motion to violent motion; it postulated three basic types of natural motion; and it attributed each type to one or more of the basic elements: circular motion around the center of the universe was attributed to aether; straight motion away from the center of the universe was ascribed to the elements air and fire; and straight motion toward the center of the universe was given to the elements earth and water. Thus, unlike natural circular motion, which can last forever, straight natural motion, especially straight-downwards, cannot be everlasting since, once the center (of the universe) is reached, the body will no longer have any natural tendency to move. Now, the terrestrial globe on which we live is essentially the collection of all things made of the elements earth and water, which have collected at the center (of the universe) or as close to it as possible. Therefore, this whole collection cannot move around the center (in an orbital revolution as Copernicus would have it), because such a motion would be unnatural, could not last forever, and would in any case be overcome by the tendency to move naturally in a straight line toward the center; further, for the same reasons, once at the center, the whole collection could not even acquire any axial rotation.

The Copernican system was also deemed physically impossible because it was in direct violation of the principle according to which *every simple body can have one and only one natural motion*. This principle was another aspect of the laws of motion of Aristotelian physics, whereas Copernicanism seemed to attribute to the Earth at least three natural motions: the revolution of the whole Earth in an orbit around the Sun, the rotation of the Earth around its own axis, and the downwards motion of parts of the Earth in free fall.

Just as the last two objections are essentially unanswerable as long as one accepts the two principles of traditional physics just mentioned, they are easily answerable by rejecting these two principles. However, rejecting them is easier said than done since, to be effective, the rejection should be accompanied by the formulation of some alternatives. In short, what was really required was the construction of a new science of motion, a new physics, which Copernicus did not provide. In fact, the alternatives were such cornerstones of modern physics as the law of inertia, the law of gravitational force, and the law of conservation of (linear and angular) momentum. For example, according to the law of inertia, the natural motion of all bodies is uniform and rectilinear; and according to the law of gravitation, all bodies attract each other with a force that makes them accelerate toward each other, or diverge from their natural inertial motion in a measurable way. Thus, the Earth's orbital motion becomes a forced motion under the influence of the Sun's gravitational attraction; the axial rotation of the whole Earth becomes a type of natural motion in accordance with conservation laws; and the downward fall of heavy bodies near the Earth's surface becomes a forced motion under the influence of the Earth's gravitational attraction.

Finally, there were theological and religious objections. One of these appealed to the authority of the Bible and may be labeled the *biblical or scriptural objection*. It claimed that the idea of the Earth moving is heretical or at least erroneous because it conflicts with many biblical passages which state or imply that the Earth stands still. For example, Psalm 104:5 says that the Lord "laid the foundations of the earth, that it should not be removed for ever"; and this seems to say rather explicitly that the Earth is motionless. Other passages were less explicit, but they seemed to attribute motion to the Sun, and thus to presuppose the geostatic system. For example, Ecclesiastes 1:5 states that "the sun also riseth, and the sun goeth down, and hasteth to the place where he ariseth." And Joshua 10:12–13 describes the following miracle, which presupposes that the Sun (not the Earth) normally moves: "Then spake Joshua to the Lord in the day when the Lord delivered up the Amorites before the children of Israel, and he said in the sight of Israel, 'Sun, stand thou still upon Gibeon; and thou, Moon, in the valley of Ajalon'. And the sun stood still, and the moon staid, until the people had avenged themselves upon their enemies."

The biblical objection had greater appeal to those (like Protestants) who took a literal interpretation of the Bible more seriously. However, for those (like Catholics) less inclined in this direction, the same conclusion could be reinforced by appeal to the *consensus of Church Fathers*; these were the saints, theologians, and churchmen who had played an influential and formative role in the establishment and development of Christianity. The argument claimed that all Church Fathers were unanimous in interpreting relevant biblical passages (such as those just mentioned) in accordance with the geostatic view; therefore, the geostatic system is binding on all believers, and to claim otherwise (as Copernicus did) is erroneous or heretical.

A third theological-sounding objection was based crucially on the idea that God is all-powerful, and it may be labeled the *divine-omnipotence argument*.[1] One of its most famous proponents was Pope Urban VIII; it was, in fact, his favorite anti-Copernican objection. A version of the argument is stated, without criticism, at the end of Galileo's *Dialogue*, and this formulation got him into trouble with Church authorities and played a role in the trial of 1633, as we shall see later.

One version of the argument claimed that since God is all-powerful, he could have created any one of a number of worlds, for example one in which the Earth is motionless; therefore, regardless of how much evidence there is supporting the Earth's motion, we can never assert that this must be so, for that would be to want to limit God's power to do otherwise. Another version seemed to argue that divine omnipotence implies that God could have created a world in which the evidence suggests a moving Earth despite its being motionless. The argument is not purely theological, but also raises issues of a logical, methodological, and epistemological nature. Moreover, some versions of the argument may very well be essentially correct and unanswerable; for example, part of the argument seems to suggest that scientific knowledge of the physical world is contingently true rather than necessarily true. On the other hand, the argument was also taken to suggest a general skeptical doubt about physical theories, as well as a specific difficulty for the Copernican theory; and these suggestions are controversial and questionable.

In summary, the idea updated by Copernicus was vulnerable to a host of counter-arguments and to considerable counter-evidence. The Earth's motion seemed epistemologically absurd because it flatly contradicted direct sense experience, and thus undermined the normal procedure in the search for truth. It seemed empirically and astronomically untrue because it had astronomical consequences that were not seen to happen. It seemed a physical impossibility because it was thought to have consequences that contradicted the most incontrovertible mechanical phenomena, and because it directly violated many of the most basic principles of the available physics. And it seemed religiously heretical, erroneous, or suspect because it conflicted with the words of the Bible, with the biblical interpretations of the Church Fathers, and with the basic theological idea of an omnipotent God.

Copernicus was aware of many of these difficulties. He realized that his novel argument did not conclusively prove the Earth's motion, and that there were many counter-arguments of apparently greater strength. Although his motivation was complex and is not yet completely understood and continues to be the subject of serious research, his awareness of the counter-arguments was an important reason why he delayed the publication of his book until he was almost on his deathbed.

Responses to Copernicanism

In light of the many objections, a common response to Copernicanism was to regard the Earth's motion as a mere instrument of mathematical calculation and observational prediction, rather than a description of physical reality. This may be labeled the *instrumentalist* interpretation of Copernicanism, and was popularized by an anonymous foreword preceding Copernicus's own preface in the printed *Revolutions*. This foreword was written and inserted without his approval or knowledge by one of the editors supervising the book's publication—Andreas Osiander. It is unlikely that Copernicus would have endorsed this interpretation since it is clear from the book that, although he was aware of the difficulties, he treated the Earth's motion as a description of physical reality (capable of being true and false), and not as a mere instrument of calculation and prediction (limited to being more or less convenient). In short, Copernicus subscribed to a *realist* interpretation of the geokinetic theory (in accordance with the doctrine of epistemological *realism*), and not to an *instrumentalist* interpretation (in accordance with epistemological *instrumentalism*).

The fact that Osiander's foreword had not been authorized by Copernicus soon became public knowledge among experts, but many scholars adopted the instrumentalist interpretation as the only way out of the difficulties.

One different response was that of Danish astronomer Tycho Brahe (1546–1601). He decided to collect new data by means of systematic naked-eye observations and the construction of new instruments. The scope, range, accuracy, and precision of his observations were unprecedented. Partly on the basis of his observational data, Tycho constructed a new theory different from both the Ptolemaic and the Copernican ones. In the Tychonic system, the Earth was still motionless at the center of the universe; the stellar sphere still had the westward diurnal motion around the Earth; and the Sun still had the eastward annual motion around the Earth. But the other planets revolved in orbits centered at the Sun, so that the system was to that extent heliocentric; but the Sun carried the whole Solar System around the motionless Earth. Moreover, Tycho did away with the solid spheres that, in some other versions of the geostatic system, carried the planets in their orbits; they no longer fit properly in the new arrangement, since some of the orbits of the heavenly bodies intersected, so that the spheres would have had to interpenetrate one another.

On a different note, the Italian philosopher and theologian Giordano Bruno (1548–1600) undertook a multifaceted defense of Copernicanism that addressed epistemological, metaphysical, theological, and empirical issues. Bruno's defense was in some ways similar to, but in important ways different from, Galileo's; it embodied more of a muddled confusion than a proper synthesis of such issues. In any case, for complex reasons, Bruno's contribution remained largely unknown, disregarded, or unappreciated; not the least of these reasons was the fact that he was burned at the stake by the Inquisition, after a long trial for heresy, lasting 8 years, during which he behaved in a defiant manner.[2]

One of Tycho's assistants inherited his data, and analyzed them in a deeper and more systematic and sophisticated manner. He was the German mathematician and astronomer Johannes Kepler (1571–1630). He rejected Tycho's compromise and was committed to the Copernican system, in part for aesthetic and metaphysical reasons. But Kepler also had a strong empirical orientation, and so he spent his life analyzing Tycho's observational data. The result was the strengthening of some key Copernican theses, such as the Earth's motion, and the refinement or revision of others. In fact, Kepler discovered that the planets revolve around the Sun in elliptical rather than circular orbits, with the Sun located at one of the two foci of these ellipses.

Unfortunately, Galileo never did pay the proper attention to Kepler's writings, and so ignored or neglected the elliptical nature of planetary orbits; the reasons for this neglect remain unclear or controversial, but there is no question that Galileo was turned off by the metaphysical flavor of Kepler's thought.[3] Nor did Galileo think much of the instrumentalist interpretation of the Copernican theory. Still less did he find acceptable Tycho's compromise of combining geostatic and heliocentric elements. However, Galileo did devise his own response to the Copernican controversy, and that is our main focus in this book. And we will explore not only how he responded, but why he responded the way he did. His intellectual motivation is at least as important as his behavior. To begin with, we should consider his early stance toward Copernicanism.

Galileo's Indirect Pursuit of Copernicanism

Galileo's earliest reference to Copernicus is found in a work entitled *On Motion*, written probably in the period 1589–92 while he was a professor at the University of Pisa, but left unpublished by him. The context is one in which Galileo is arguing, against Aristotle, that when the motion of a body changes from one direction into the opposite direction, there does not have to be a state of rest at the turning point. One of Galileo's several arguments is that oscillatory motion along a straight line can be generated by combining two continuous circular motions: consider two equal circles such that the center of each is located on the circumference of the other, and such that they rotate in opposite directions at different rates; then one can adjust these rates of rotation so that there is a point on the circumference of the faster rotating circle which moves back and forth along a straight line (as seen from a location outside both circles). Galileo explicitly credits Copernicus's work *On the Revolutions* as the source of this demonstration.[4]

This reference is important for two reasons. First, it shows that Galileo was intimately acquainted with Copernicus's masterwork at the beginning of his career. Second, it is even more important for what Galileo does *not* say: Copernican astronomy is nowhere in sight, and so he must not have been impressed by the cogency or conclusiveness of Copernicus's arguments for the geokinetic thesis. Other passages in Galileo's *On Motion* do have a connection with the Earth's rotation, but that connection is indirect and to appreciate it we must wait for other clues, which will be elaborated presently.

Galileo's first explicit discussion of Copernican astronomy is found in a letter he wrote to Jacopo Mazzoni, dated May 30, 1597. Mazzoni had been a senior colleague and good friend of Galileo's during his time at the University of Pisa. Mazzoni was an eclectic philosopher with wide interests, who held anti-Aristotelian ideas on the motion of falling bodies and their speed of fall; and these ideas overlapped with Galileo's own. In 1597, Mazzoni had just published a book containing a critical comparison of Plato and Aristotle. Galileo, who was then at the University of Padua, had just read the book and was writing to congratulate the old friend and to express his gratification at the fact that they seemed to agree about many things. However, the book also contained an anti-Copernican argument, and most of Galileo's letter is a lengthy analysis and refutation of that argument.

Mazzoni had argued as follows. If the Earth revolves around the Sun, then this off-center location would imply that terrestrial observers would not always see exactly half of the stellar sphere; rather, they would see less than half at midnight and more than half at noon. Because of the immense size of the stellar sphere, the difference would be very small; but it should be noticeable, because on the Earth by climbing a tall mountain (such as Mount Caucasus) one can notice a comparable difference in the visual horizon. However, we always see exactly half of the stellar sphere. It follows that the Earth is not located off-center revolving around the Sun.

Galileo's refutation is the following. If the Earth revolves around the Sun, the difference in visibility of the stellar sphere between midnight and noon would be equal to that caused on the Earth by climbing a mountain whose height is 1 and 1/7 miles. On the Earth, the difference in visibility of the stellar sphere resulting from climbing such a mountain is 1 degree and 32 minutes on each side. These quantities are based on the traditional estimates of astronomical distances, which are: distance between Earth and the Sun = 1216 Earth radii; radius of the stellar sphere = 45,225 Earth radii; and Earth radius = 3035 miles. However, the Copernican estimate of the size of the stellar sphere is much greater. Thus, the difference in stellar horizon would be much less than 1 degree and 32 minutes; and that would be unlikely to be noticeable.

The most relevant and important aspect of Galileo's letter to Mazzoni is that it constitutes an explicit defense of Copernicanism from an astronomical objection. Moreover, the core of Galileo's reasoning is mathematical or quantitative, and so what we have here is a mathematical defense. Combining these two aspects, we might say that Galileo is exhibiting a mathematical appreciation of Copernicanism, and this is in accordance with his earlier reference to Copernicus in *On Motion*, discussed earlier. However, it is unclear that we can describe Galileo's attitude any more precisely than is conveyed by the vague notion of what I am labeling "appreciation."

In fact, Galileo described his stance in a similarly vague and unclear manner in the introductory part of this long letter. There, he reminded Mazzoni that in the first years of their friendship (1589–92) they often engaged in amicable debates on astronomical topics, and that, for the sake of the argument, he (Galileo) would then take the Copernican side. However, now (in 1597) Galileo had some feelings towards the topic of the Earth's motion and location, which he did not have earlier.

More clues about Galileo's attitude toward Copernicanism are found in a letter he wrote to Kepler a few months thereafter (August 4, 1597). The letter was occasioned by the fact that Galileo had just received a copy of Kepler's book *The Secret of the Universe*, published the previous year, and he wanted to thank Kepler. Galileo said that he had only read the introduction, but planned to read the rest. Then he added several clarifications about his own stance toward Copernicanism.

Galileo indicates that he is engaged in a program of physical research that fits well with Copernicanism, but not at all with the Aristotelian Ptolemaic view. We might say that he is pursuing the physical side of Copernicanism, physical by contrast with Copernicus's own astronomical motivation, or with the metaphysical flavor of Kepler's own book. Galileo is obviously referring to the sort of theory of motion which he had been working on for some time and part of which is recorded in his work *On Motion*. In particular, he claims to have found that many physical phenomena cannot be explained by the geostatic theory, but are explicable on the basis of the geokinetic hypothesis. He does not explicitly mention the phenomenon of the tides, but there can be no doubt that he had the tides in mind.

Another clarification is this. Galileo explicitly says that he is in possession not only of some positive evidence or constructive reasons favoring Copernicanism, but also of criticisms or refutations of counter-arguments and objections. Thus, he is implicitly suggesting that the negative aspect of the investigation is also essential, namely that the elaboration of Copernicanism must include a serious and careful critical component. In fact, a few months earlier, in his letter to Mazzoni, Galileo had just conducted one exercise in such a critical defense. However, he is clear that this is just one example of many. Indeed, if one examines the historical context, one can identify these other anti-Copernican arguments to be the mechanical objections to the Earth's motion that had recently been advanced by Tycho Brahe, in books published in 1588 and 1596.

Finally, it is very revealing that Galileo expresses a fear to publish his Copernican inclinations and pursuits. This expression is an explicit comment on his view of the strength of the arguments in support of

Copernicanism. He obviously does not think that the Copernican arguments are conclusive or even strong enough to convince someone who, unlike Kepler, is not already favorably inclined.

Now that we have a general description of the type and of the strength of Galileo's stance and rationale, let us see whether we can identify some of the constructive arguments more precisely. I have already mentioned that one of them was probably the tidal argument; but, since he did not elaborate it until later, we shall postpone discussion of it. Instead, let us see whether we can identify any others that were demonstrably in his mind in the earlier period. I believe one can be found in the section dealing with the Earth being motionless of his *Treatise on the Sphere, or Cosmography*.

It is well known that this *Cosmography*, which Galileo never published, was not meant to be an original contribution to knowledge, but a concise and elementary introduction to spherical astronomy for beginning students; and it is likely that he used it in both his university courses and his private tutoring. Moreover, it cannot be denied that the work is generally conservative, Aristotelian, and Ptolemaic in content. In regard to its form, however, it has some original and interesting elements. One is the methodological and epistemological introduction, where Galileo explicitly discusses the method of hypothesis, giving the following examples of hypothetical assumptions: that the sky is spherical, that it moves circularly, that the Earth stands still, and that it is located at the center. The other aspect is the impersonal, informational, and noncommittal style which Galileo uses as he discusses the details of traditional astronomy. The Copernican system is explicitly mentioned only once, and this brings us to the passage referred to earlier.

Unlike other sections, this one, entitled "That the Earth Is Motionless," begins with an admission that this is a controversial question. However, he is quick to add that nevertheless he is primarily going to present some arguments for the geostatic thesis of Aristotle and Ptolemy. Then he goes on to argue why the Earth cannot have rectilinear motion. After that he takes up the question of its possible rotation, concerning which we have the following very important passage:

> But, that it may move circularly has more verisimilitude, and therefore some have believed it; they have been moved principally by their considering it almost impossible that the whole universe except the earth should experience a rotation from east to west in the period of 24 hours, and hence they have believed that it is rather the earth which undergoes a rotation from west to east during such a time.[5]

He does not say whether or why he rejects this argument. Instead, in his usual impersonal style he goes on to summarize the traditional objections from falling bodies, birds, clouds, ship's mast experiment, and whirling.

The argument in the passage quoted above is important because it is obviously an appeal to Galileo's own doctrine of natural and neutral motions, which he had elaborated at least as early as 1589–92 in his work *On Motion*.[6] This is the theory according to which there are three basic types of motions: natural motion, or motion where the object approaches its natural place; violent motion, or motion where the object recedes from its natural place; and neutral motion, or motion which is neither natural nor violent. For Galileo, examples of neutral motions would be the rotation of a homogeneous sphere at the center of the universe or the rotation of a sphere around its center of gravity even if the sphere is located elsewhere. Finally, he thinks that to start a body moving with neutral motion, a force as small as you like is sufficient. Given this doctrine, it presumably follows that the Earth would have axial rotation even in an otherwise Ptolemaic universe.

Galileo explicitly recognized this type of argument as one favorable to Copernicus both in the *History and Demonstrations Concerning Sunspots*, where he argued in support of solar rotation, and in the *Dialogue*, where he gave it at the beginning of the Second Day in support of terrestrial rotation.[7] So it is likely that he had made the connection even in the pre-1609 period.

Finally, some items of negative evidence are also revealing of Galileo's early stance toward Copernicanism. On May 4, 1600, Tycho wrote a letter to Galileo, but he never answered it. Moreover, Galileo also did not answer the letter dated October 13, 1597 which Kepler wrote after receiving his. Kepler was asking to be informed of Galileo's Copernican arguments and wanted him to make certain observations to try to detect stellar parallax. There were undoubtedly external factors that contributed to Galileo's lack of cooperation, such as Kepler being a "heretic," namely a Protestant. Moreover, as already mentioned, Galileo found Kepler's approach too metaphysical. And here we can add a more internal and methodological reason: Galileo's lack of interest in pursuing the astronomical side of Copernicanism, and his dissatisfaction with the strength of the pro-Copernican arguments.

We may summarize the pre-telescopic period by saying that Galileo's attitude toward Copernicanism was one of partial, qualified, and indirect pursuit. Such pursuit was based largely on its compatibility with his own new physics of motion. However, he neither believed nor accepted Copernicanism as true. Indeed, as he confessed later, he was much more impressed by the observational astronomical objections against it, and deemed them to be strong and unanswerable.

CHAPTER 4

RE-ASSESSING COPERNICANISM (1609–1616)

The Effect of the Telescopic Discoveries

In 1609, Galileo heard about an optical instrument invented in Holland the year before, consisting of an arrangement of lenses that magnified images three to four times. Without having a prototype in his possession, he was soon able to duplicate the instrument, mostly by trial and error (see Figure 9). He was also able to increase its magnifying power first to 9, then to 20, and, by the end of the year, to 30. Moreover, rather than merely exploiting the instrument for practical applications on Earth, he started using it for systematic observations of the heavens, to learn new truths about the universe.

Within three years, Galileo made several startling discoveries. The Moon had a rough surface full of mountains and valleys, similar to land on the Earth (Figure 10). Innumerable other stars existed besides those visible with the naked eye. The Milky Way and the nebulas were dense collections of large numbers of individual stars. The planet Jupiter had four moons revolving around it at different distances and with different periods. The appearance of the planet Venus, in the course of its orbital revolution, changed regularly in a manner analogous to the phases of the Moon: from a full disc, to half a disc, to crescent, and back to a half and a full disc (Figure 11[1]). And the surface of the Sun was dotted with dark spots that were generated and dissipated in a very irregular fashion and had highly irregular sizes and shapes, like clouds on Earth; but that while they lasted, these spots moved regularly in such a way as to imply that the Sun rotated on its axis with a period of about one month (Figure 12).

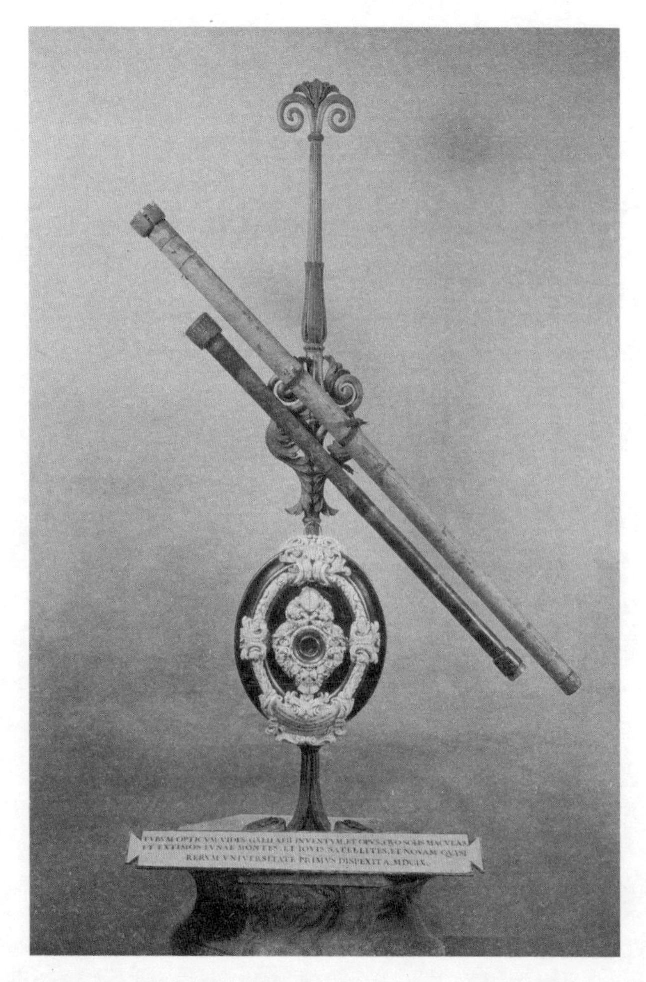

Figure 9. Galileo's original telescopes, in a previous display at Museo Galileo, Florence

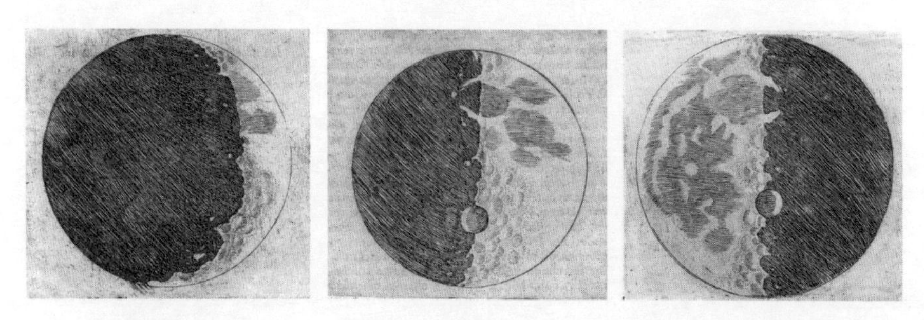

Figure 10. Appearance of the Moon through Galileo's telescope

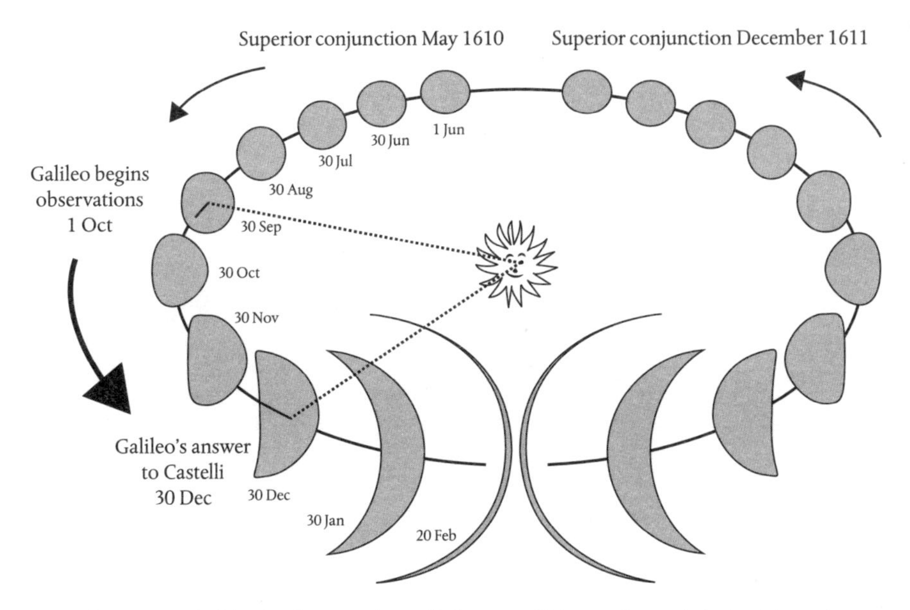

Figure 11. Galileo's observation of phases of Venus, from Palmieri (2001)

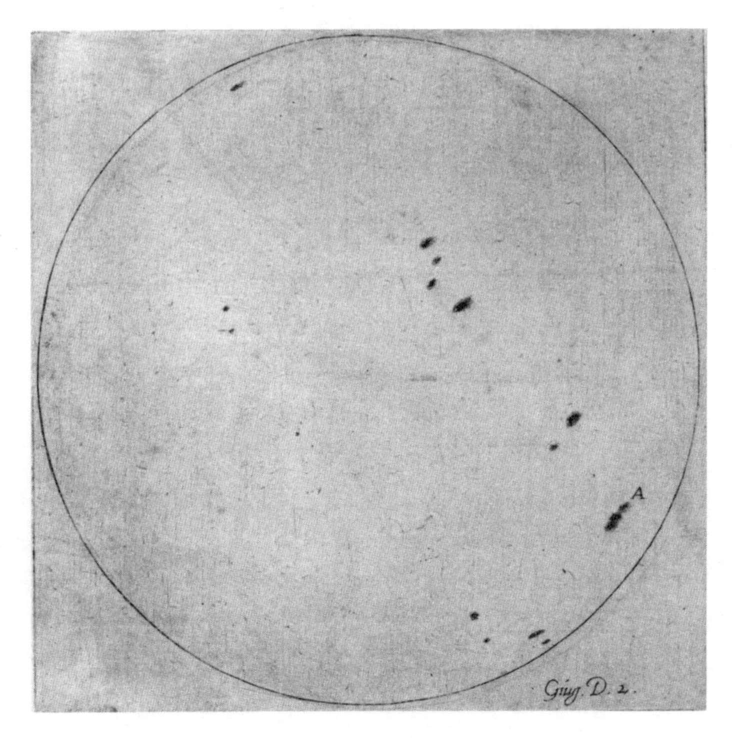

Figure 12. Sunspots as observed by Galileo on June 2, 1612

There is no question that the telescopic discoveries led Galileo to a significant reappraisal of Copernicanism. Less easy to ascertain is what the change was exactly, how sudden or slow it was, and what the precise motivating reasons were. The first piece of significant evidence here is the book he was quick to publish in 1610, *The Sidereal Messenger*. There are three relevant passages in this work.

One occurs in the author's dedication to the Grand Duke of Tuscany. Since this is often taken as the first published evidence that Galileo accepted Copernicanism, it is important to be especially careful. At this point in the dedication Galileo is calling attention to the satellites of Jupiter and to the fact that he had named them Medicean stars, after the ruling family of Tuscany. The paragraph reads as follows:

> Behold then, reserved for your famous name four stars, belonging not to the ordinary and less-distinguished multitude of the fixed stars, but to the illustrious order of the wandering stars; like genuine children of Jupiter, they accomplish their orbital revolutions around this most noble star with mutually unequal motions and with marvelous speed, and at the same time all together in common accord they also complete every twelve years great revolutions around the center of the world, certainly around the sun itself.[2]

The final clause of this passage is the expression often taken as evidence of commitment to Copernicanism. But that is done by translating the original Latin to read, or by interpreting it to mean "around the center of the world, that is the sun."[3] However, I believe it is more correct to have it the way I rendered it above, which embodies a certain ambiguity. That is, a body that follows Jupiter in its orbit is likewise encircling the smaller orbit of the Sun, and in that sense is moving around the Sun; and this is certain, with both sides of the controversy agreeing. But the clause could also mean that Jupiter and its satellites move around the Sun as center, and then we would have some acceptance of Copernicanism. At any rate, this would involve directly only the heliocentricity of planetary motions, and not necessarily the other elements of the system (such as the Earth's motion). So, although there is no question that Galileo is expressing a favorable attitude toward

Copernicanism, and that he is now involved with its astronomical aspects, any more precise definition of his stance is elusive.

This conclusion corresponds very well with the actual content of *The Sidereal Messenger*, in which we find only two places where the status of Copernicanism is discussed explicitly. They both involve rebuttals of traditional anti-Copernican objections.

In one passage in the middle of the book, Galileo indicates he is now in the position of being able to answer the objection that the Earth cannot be a planet because it is devoid of inherent light; his telescopic discoveries about the optical properties of the Moon enable him to say that the Earth is not essentially different from the Moon in this respect. More generally, he thinks his lunar discoveries are such "that the connection and resemblance between the moon and the earth may appear more plainly";[4] that is, these discoveries show the untenability of the Earth–heaven dichotomy, and so undermine the anti-Copernican objection from the Earth–heaven dichotomy.

The other passage occurs at the end of the book. There, Galileo explains that the ability of Jupiter's satellites to keep up with Jupiter as it revolves in its orbit allows one to refute the traditional lunar-orbit objection. This was the argument that the Earth cannot revolve around the Sun because the Moon clearly orbits the Earth, and so would be left behind.

Of course, these refutations of objections do not prove Copernicanism; they merely strengthen it. The attitude expressed by Galileo may be described as indicating a more direct, active, and explicit kind of pursuit, and a higher degree of favorable appraisal than before.

Finally, there is an important piece of negative evidence in *The Sidereal Messenger*, regarding something which Galileo does *not* say or do. In the printed book he dropped a clause which he had written down in the manuscript draft. This occurs at the end of the discussion of Jupiter's satellites, near the end of the book. The manuscript draft speaks of "the Copernican system (which above all I judge to be consonant with the truth),"[5] but the printed book lacks the parenthetical remark. Clearly, in the printed book, Galileo was being more cautious in his endorsement of Copernicanism.

The next significant document for this period is Galileo's letter to Giuliano de' Medici, Tuscan ambassador to Prague, dated January 1, 1611. Its main purpose was to decipher the anagram Galileo had sent him in an earlier letter. When properly transposed the anagram stated that Venus shows phases like the Moon, a phenomenon Galileo had been able to observe with the help of the telescope. The attitude he displays now is not, as some scholars have alleged, complete acceptance of the Copernican system, but rather acceptance of two specific theses in it. In the text this is as clear as his emphasis on empirical accuracy:

> From this marvelous observation we have sensible and certain demonstration of two great questions, which so far have been debated by the greatest minds of the world: one is that planets are all dark (since the same thing happens with Mercury as with Venus); the other is that Venus necessarily revolves around the sun, as do also Mercury and all other planets.[6]

He goes on to describe the change in his attitude as one from belief without, to belief with, empirical proof: "This had indeed been believed by the Pythagoreans, Copernicus, Kepler, and myself, but not sensibly proved, as done now for Venus and Mercury."[7] This change should not be equated with that from pursuit to acceptance; rather it is a change from one kind of acceptance to another, the difference being the grounds for the acceptance; in other words, we have a development from a situation in which acceptance was based on factors other than empirical adequacy, to a situation in which it was based on empirical evidence. On the other hand, non-empirical acceptance should be distinguished from an attitude of nonbelief, pursuit, or exploration. In fact, Galileo makes the latter distinction when he goes on to congratulate Kepler and other Copernicans for their pre-telescopic intuitions, but does not include himself: "Thus, Mr. Kepler and the other Copernicans will have reason to be proud for their having believed and philosophized correctly."[8]

The next important milestone in our story is a letter to Prince Federico Cesi, a wealthy aristocrat with an amateur interest in science; he was the founder and head of the Lincean Academy, the first international scientific society, into which Galileo himself was inducted in 1611. On June 20, 1612, Cesi had written to Galileo, expressing his attraction to Copernicanism because of its doing away with epicycles and eccentrics; he also asked Galileo's opinion on the problem that these seemed unavoidable for the case of the terrestrial and lunar orbits, given the periodic changes in distance between the Earth and the Sun, and between the Earth and the Moon.

Ten days later Galileo replied that "we must not desire that nature should accommodate herself to what seems better arranged and ordered to us; rather it is appropriate that we should accommodate our intellect to what she has done, certain that this and nothing else is the best."[9] He then goes on to apply this principle by arguing that if by epicycles we mean orbits not encompassing the Earth, then we must admit their reality, examples being the revolutions of Jupiter's satellites around that planet, and the orbits of Venus and Mercury around the Sun; and if eccentrics are meant relative to the Earth, then the orbit of Mars encompassing the Earth is a clear example, since the telescope reveals that its apparent magnitude is 60 times greater at certain places of its orbit than at others.

Galileo is here explicitly expressing some theoretical limitations to the principle of simplicity, as well as practically utilizing it in a particular case. This is in accordance with his behavior during the pre-telescopic period. At that time, as we have seen, Galileo did not attach enough weight to the greater simplicity of the Copernican system either to accept it or to ground his pursuit on that. This is not to say that he completely disregarded the criterion of simplicity; only that he did not attach significant or decisive weight to it.

Let us now turn to the *History and Demonstrations Concerning Sunspots*, which was written in 1612 and published in 1613. This work is widely recognized as containing the strongest endorsement of Copernicanism that Galileo ever published, either before or after this date. What is less well known and seldom discussed is the exact form and context that this endorsement takes. It occurs at the very end, in a passage written on December 1, 1612.

There, Galileo finishes with the topic of sunspots about three pages from the end, by stating his main conclusion that the spots are on the Sun and made of some kind of volatile substances like clouds on Earth. Then he goes on to report some observations about the planet Saturn. Galileo was actually seeing what we now know to be rings around this planet, although he was never able to formulate this correct interpretation. When he first observed them in 1610, he thought he was seeing Saturn as consisting of three distinct but interconnected bodies. However, in this passage, he now reports that Saturn no longer appeared as three-bodied.

Galileo admits his puzzlement and confusion, but goes on to make several very daring and rather precise predictions about the periodic reappearance and re-disappearance of Saturn's "companions," up to the summer of 1615. He does not, however, reveal the conjecture on which he says he based these bold predictions, but promises to do so later, after events would confirm or disconfirm him. He concludes his discussion as follows:

> As few doubts as I have about their return, I am proceeding with restraint in regard to their particular features, for these are based at present on probable conjecture alone. But whether these things take place precisely in this fashion or in another, I say to Your Lordship that this star, too, and perhaps no less than the emergence of horned Venus, agrees in a wondrous manner with the harmony of the great Copernican system, to whose universal revelation we see such favorable breezes and bright escorts directing us, that we now have little to fear from darkness and cross-winds.[10]

Two main claims in the last part of this quotation are worth noting. The first is that *perhaps* Saturn's behavior too confirms Copernicanism. Despite the fact that Galileo never did reveal the theoretical basis of his prediction and the connection with Copernicanism, and despite the fact that that basis remains a puzzle for scholars, the judgment is obviously based on the criterion of empirical accuracy.

The second claim is that now he thinks all evidence is pointing toward Copernicanism and seems to have little doubt about its correctness. Although Galileo does not explicitly include sunspots in this evidence, the connection is obvious enough that it can easily be attributed to him. For as he will argue later in the *Dialogue*, sunspots contribute to the empirical undermining of the Earth–heaven dichotomy, and thus to the strengthening of

Copernicanism. Yet, the *Sunspots* book does *not* contain one of the most powerful pro-Copernican arguments—the argument based on the annual cycle of sunspot paths, which, as we shall see (Chapter 6), is persuasively advanced in the *Dialogue*. Although at this time (1612–13) Galileo was in a position to observe the annual motion of sunspots, and to formulate the geokinetic explanation of that phenomenon and the corresponding argument, there is no evidence that he did either until 1629–31, while in the process of writing the *Dialogue*.

In the Postscript that was added to the published version of the *Sunspots* book, there is another important clue that at about this time Galileo was finding another piece of evidence which could be explained only on the basis of the Earth's orbital revolution. The phenomenon involved the eclipsing of Jupiter's satellites and the variations in the duration of these eclipses. The details of the argument were never written up by Galileo, and they are extremely technical. Scholars have tried to identify the relevant documentary evidence and to piece together the main points.[11] However, Galileo's claim in the Postscript is as clear and unambiguous as one could wish, namely that such eclipsing is evidence for the Earth's annual orbital revolution.

The *Sunspots* book also contains two other passages that are relevant to understanding Galileo's re-assessment of Copernicanism. One is the theory that the Sun rotates on its axis, which represents a modification (however slight) of Copernicus's original system. The other is a criticism of an argument given by Jesuit astronomer Christoph Scheiner to show that Venus revolves around the Sun; of course, Galileo accepts this conclusion, but he finds several faults with Scheiner's attempt to ground it on the alleged observation of a transit of Venus across the solar disk, rather than on the phases.[12]

Both passages show a piecemeal attitude that might be called methodological gradualism, as contrasted to methodological holism. That is, Galileo does not hesitate to modify various elements of the Copernican system, or to reject proposed contributions by other supporters, as the case requires. This suggests, it seems, a greater commitment to certain procedures than to any specific physical or cosmological theses; for the case of solar rotation the procedure in question would seem to be related to empirical accuracy, while for Scheiner's argument about Venus it would seem to involve correct reasoning.

Galileo's re-assessment of Copernicanism comes to a climax in 1614, as we can see in his letter to Giovanni Battista Baliani dated March 12. Baliani was a government official of the Republic of Genoa who was engaged in serious research on the physics of falling bodies, and his contributions overlapped somewhat with Galileo's. In this letter, we find for the first time a genuine expression of certainty, together with a summary of his reasons, as well as a reasoned rejection of Tycho's theory. The crucial passage is the following:

> As regards Copernicus's opinion, I really hold it as certain, and not only because of the observations of Venus, of sunspots, and of the Medicean planets, but because of his other reasons, and because of many other particular reasons of mine, which seem conclusive to me... In Tycho's opinion I still find all the very great difficulties which make me abandon Ptolemy, whereas in Copernicus I find nothing which gives me the least scruple, and least of all the objections which Tycho makes to the earth's motion in certain letters of his.[13]

There is no question that we have here an endorsement of Copernicanism stronger than any we have seen up to this date, and stronger than we find anywhere else subsequently. And it is clear that Galileo uses labels that are epistemically loaded: "certain" to characterize the position, and "conclusive" to describe the supporting arguments. Nevertheless, I do not think we have here a qualitative jump from what came immediately before; it is a change of degree. Besides, the notions of certainty and conclusiveness are used loosely in this kind of informal context. And in any case Galileo's words must be balanced against other considerations.

For a start, this is a private letter, and there is no analogous expression in any of Galileo's published works. Also, Galileo was *not* at this time writing the work on the "System of the World" which he had promised in *The Sidereal Messenger*. Although there were many causes for this delay, including external ones, one contributing factor may have been that he did not have yet all the arguments and evidence required to be really sure, as he was claiming in this letter; so the crucial words here are perhaps merely that—words.

And there is another, more complex consideration. This alleged certainty and conclusiveness may be no more literally true than is Galileo's assertion in the same passage that he finds nothing in Copernicus which gives him "the least scruple." In fact, Galileo had several qualms about Copernicanism. One scruple was that, as we have just seen, solar rotation represents a departure (however minor) from Copernicus. Moreover, and more significantly, Galileo never accepted Copernicus's "third motion," according to which the terrestrial axis was supposed to "precess" (wobble) with an annual period in order to compensate for the orbital revolution and thus enable the axis to remain always parallel to itself; instead, as the later *Dialogue* clearly and plausibly explains, the constant parallelism of the Earth's axis is an instance of rest (or inertia, as we might say). Another qualm was that, as the *Dialogue* also shows (and as we shall see later), Galileo never refuted the objection from stellar parallax, but rather accepted the difficulty, and suggested a way of testing for it.

Despite all these provisos, the endorsement neither can nor should be ignored. So let us look at the motivating reasons. Notice that Galileo speaks of three groups of reasons: first, the observational ones depending on the telescope and involving the phases of Venus, sunspots, and Jupiter's satellites; second, Copernicus's own reasons; and third, Galileo's own "other particular reasons." The first group were the chronologically latest ones which had been accumulating in the past several years since the telescope, and which seem to have caused a qualitative change in Galileo's attitude, from something like indirect pursuit or qualified rejection to something like direct pursuit or qualified acceptance. These reasons relate intimately to the principle of empirical accuracy.

The second group (Copernicus's own reasons) were never, not even in the *Dialogue*, discussed by Galileo in any great detail. We do know, as we saw above, that for Galileo they were not primarily simplicity considerations, at least not in any simple sense of the notion of simplicity. Instead, I believe that they are best seen as considerations of explanatory coherence or against *ad-hocness*. Galileo was sensitive to the importance of explanatory coherence, however insufficient he may have regarded it as a basis of acceptance, or even as a sole basis of pursuit. And in view of Galileo's qualms about the naïve application of the principle of simplicity, notice that this conception of Copernicus's reasons is also different from the simplicity interpretation.

Finally, Galileo also had a third group of reasons, and these he had had for a very long time. Collectively, they reduced, as we have seen already, to the fruitfulness of his theory of motion, a theory quite compatible with Copernicanism, and quite at odds with the Aristotelian–Ptolemaic system. Specifically, they included not only the argument from neutral motion (discussed in Chapter 3), but also the argument based on a geokinetic explanation of the tides (which we will consider later). Since Galileo claims he has "many" such arguments, one wonders what else he had in mind.

One clue to what they might have been lies in Galileo's remark about the Tychonic system. He says he has the same basic difficulties with it as with the Ptolemaic system. These can only be problems stemming from his theory of motion, or as we would say, dynamical difficulties. They are sketched and summarized at the beginning of the Second Day of the *Dialogue*. Two points are especially pertinent here. First, these objections apply with equal force to both the Tychonic and the Ptolemaic versions of the geostatic system, and so the often-heard criticism that Galileo is guilty of neglecting the Tychonic system is without foundation. Galileo was not insensitive to the need to appraise theories in the light of rival alternatives, and his behavior shows that he appraised them not just vis-à-vis empirical data, but also on a comparative basis. Second, those anti-geostatic objections are explicitly labeled as probable and not conclusive by Galileo, and this helps us to resolve one last question about the present passage.

Here Galileo judges the strength of his reasons as "conclusive." The actual expression and punctuation he uses ("which seem conclusive to me") is ambiguous, since the conclusiveness could be referring to his third group of reasons individually, or his third group of reasons as a whole, or all his three groups of reasons collectively. I believe that the last one is meant, as suggested by the just-mentioned passage in the *Dialogue* (beginning of the Second day). But one difficulty now remains. It is this. Since the tidal argument is obviously included in the third group, could he not here be attributing conclusiveness to the tidal argument? It is conceivable that he might, but in fact he is not. It can be argued that the tidal argument is presented in the *Dialogue* as an inductive, probable, hypothetical, non-necessitating argument, but the *Dialogue* was written after the anti-Copernican decree of 1616, so Galileo had external motives for such a presentation. We need to examine the tidal argument in the version found in his "Discourse on the Tides," written in January 1616, two months before the Decree of the Index.

An important but often overlooked fact about the "Discourse on the Tides" is that it is not just about tides, but also about winds. In other words, we have not one but two arguments for the Earth's motion, the first based on the tides, the second on prevailing easterly winds, the so-called trade winds. This immediately suggests that neither one is considered to be absolutely conclusive, for the argument from the trade winds would be superfluous if the tidal argument were conclusive, and vice versa. It is true that in the twenty pages or so of this essay, only about two at the end are devoted to the wind argument, but that only means that the topic of tides involves more details and that the argument has more complications. So if length were at all relevant, that might actually weaken it by exposing it to more potential difficulties or errors. Moreover, degrees of strength may be a function of length, but absolute conclusiveness is not because it would lead to the absurdity of labeling the tidal argument both absolutely conclusive and also nine times stronger than the wind argument.

This is reinforced by frequent explicit remarks on Galileo's part. For example, after listing several possible causes of the motion of water in general, he introduces the connection between tides and the Earth's motion with words that indicate tentativeness on his part. This tentativeness is typical of the rest of this essay, but we shall limit ourselves to a concluding note on the last page of the essay, which is simply too revealing to be ignored:

> I could propose many other considerations if I wanted to delve into finer details. Many, many more could be advanced if we had abundant, clear, and truthful empirical reports of observations made by competent and diligent men in various places of the earth; for by comparing and collating them with the assumed hypothesis we could decide more firmly and ascertain more

> correctly the things that pertain to this very obscure subject. At the moment I only claim to have given something of a sketch, suitable at least for stimulating students of nature to reflect on this new idea of mine. I hope, however, that it does not turn out to be delusive, like a dream which gives a brief image of truth followed by an immediate certainty of falsity. This I submit to the judgment of intelligent investigators.[14]

In conclusion, there is no doubt that the certainty expressed by Galileo in the letter to Baliani was an inductive and practical, rather than an absolute, kind of certainty; it was based on the practical conclusiveness of all the arguments taken together, physical, telescopic, and Copernicus's original ones. Certainly he did not think that any one individual argument or piece of evidence was absolutely conclusive.

This analysis also fits very well with another series of relevant remarks in another key document of the period, Galileo's *Letter to the Grand Duchess Christina* (1615). The main purpose of this letter, as we shall see later, was to try to defend the Copernican system from the scriptural objection; in other words, to explore whether Copernicanism is really incompatible with Scripture. However, in the letter's introductory part, to set the stage for the discussion, Galileo describes his attitude toward Copernicanism.

Galileo tells us, "in my astronomical and philosophical studies, on the question of the constitution of the world's parts, I hold that the sun is located at the center of the revolutions of the heavenly orbs and does not change place, and that the earth rotates on itself and moves around it."[15] This is a clear and explicit statement of endorsement, but the strength and nature of this endorsement must be inferred from other statements. There are two sets of relevant statements: those that are meant to clarify his relationship to Copernicus, and those that are intended to explain how Galileo's cosmological position relates to his own astronomical discoveries.

By the latter it is obvious that Galileo is referring to such things as the lunar mountains, Jupiter's satellites, the phases of Venus, and sunspots. Although these were questioned at first, he now regards their existence and main features as conclusively proved, for he notes with pride that "then it developed that the passage of time disclosed to everyone the truths I had first pointed out."[16] By contrast, about the geokinetic hypothesis, Galileo says that

> I confirm this view not only by refuting Ptolemy's and Aristotle's arguments, but also by producing many for the other side, especially some pertaining to physical effects whose causes perhaps cannot be determined in any other way, and other astronomical ones dependent on many features of the new celestial discoveries; these discoveries clearly confute the Ptolemaic system, and they agree admirably with this other position and confirm it.[17]

The key notion here is that of confirmation. He seems to regard the Copernican position as confirmed. What does this mean?

That it does not mean conclusively proved is shown by Galileo's understanding of his relationship to Copernicus. On the next page of the letter's introductory part, in the context of discrediting some of his opponents who thought that the geokinetic idea was Galileo's invention, he clarifies that "Copernicus was its author, or rather its reformer and confirmer."[18] The same terminology of confirmation is used. Thus, it is obvious that Galileo thinks he is doing more of the same of what Copernicus did. There is no claim of a breakthrough from Copernicus's mere confirmation to his own strict demonstration.

There is, of course, a strengthening of the position, which Galileo describes with the words that now "one is discovering how well-founded upon clear observations and necessary demonstrations this doctrine is."[19] He does not say that the Earth's motion is now clearly observed and necessarily demonstrated, but that it is well founded on them. The necessary demonstrations referred to must be those that prove the truth of his celestial discoveries mentioned earlier. There is no problem, of course, about a long and complex probable proof, such as that supporting the Earth's motion was for him at that time, consisting partly of segments that are necessary demonstrations or clear observations, because the final conclusion would be only as strong as the weakest supporting subargument. And Galileo clearly realizes this since, apropos of Copernicus, one remark he makes is that parts of his work too consist of clear observations and necessary demonstrations. That is, rather than getting involved in biblical interpretation, Copernicus "always limits himself to physical conclusions pertaining to celestial motions, and he treats of them with astronomical and geometrical demonstrations based above all on sense experience and very accurate observations."[20]

To summarize, in the post-telescopic period, Galileo came to regard the geokinetic theory, compared to the geostatic theory, as much better supported by astronomical, physical, and philosophical arguments and evidence. This re-assessment represented a kind of reversal of his pre-telescopic judgment. The main factor that tipped the balance was the telescopic astronomical discoveries, which provided answers to almost all astronomical objections to Copernicanism and some positive support for the Earth's motion. However, even after the telescope he was not insensitive to the existence of some unanswerable objections to Copernicanism and to the fact that it had not yet been conclusively proved, although he regarded the case in favor of the Earth's motion as very strong and increasingly more convincing, and the case for the Earth's rest as vanishingly weak. Similarly, in the pre-telescopic period, although he regarded the case for Copernicanism very weak, and the case for the geostatic system as overwhelming, he was not blind to the existence of some good pro-Copernican arguments, such as the physical arguments of his own invention.

For a complete re-assessment of Copernicanism, besides the astronomical, mechanical, and epistemological arguments, one had to examine the scriptural and theological objections. Galileo was well aware of this, but had little inclination to do so, for various reasons. However, little by little, and willy-nilly, he was dragged into the latter discussion as well, with fateful consequences.

Scripture vs. Copernicanism

Scriptural criticism of Copernicus's theory was immediate. We now know this from the censure of his *Revolutions* (1543) written by Dominican friar

Giovanni Maria Tolosani in 1546–7. Indeed, such criticism even antedated the publication of Copernicus's masterpiece, since his ideas had circulated earlier through his own unpublished writings and the publications of some followers. For example, in an incidental remark in 1539, Martin Luther criticized Copernicanism as incompatible with the biblical passage in Joshua 10:12–13.[21] An even more negative assessment is found in the preface to the second edition (1541) of a Copernican book by a friend and former student of Copernicus's named Georg Joachim Rheticus; in that preface, Rheticus included a letter by one of his own friends praising the Copernican theory, but also expressing the worry that clergymen will judge it to be heretical.

It is also well known that the criticism continued, for scriptural objections were usually included in discussions of the status of heliocentrism. However, it was not until Galileo's telescopic discoveries in 1609–12 that the problem became a crisis. I believe the key reason for this crisis was that, as we have seen, these discoveries entailed a major re-assessment of Copernicanism: they suggested that the Earth's motion and its off-center "heavenly" location could now be regarded as real possibilities and not merely convenient instruments of astronomical calculations and predictions, and indeed as more likely to be true than the alternatives, although they did not provide a conclusive demonstration of the physical truth of the Copernican theory.

Galileo's *Sidereal Messenger* left the printing press in March 1610. Three months thereafter, Martin Horky published *A Very Short Peregrination Against the Sidereal Messenger*. A few months after that, Ludovico delle Colombe compiled an essay, "Against the Earth's Motion," that included theological objections; it circulated widely, but was left unpublished. The following year, Francesco Sizzi published in Venice a book objecting on scriptural grounds to Galileo's discovery of the satellites of Jupiter. In 1612, Giulio Cesare Lagalla, professor of philosophy at the University of Rome, published a book disputing Galileo's lunar discoveries.

By the summer of that year, Galileo was worried enough that he asked Cardinal Carlo Conti for advice on whether Scripture really favors Aristotelian natural philosophy and contradicts Copernicanism. Conti was an influential churchman in Rome and replied promptly in two thoughtful letters. His views can be summarized as follows. It is the Aristotelian doctrine of heavenly unchangeability that contradicts Scripture and the common opinion of Church Fathers. However, it will take some time to determine whether the new discoveries establish heavenly changeability since, for example, some will try to explain sunspots in terms of swarms of small planets circling the Sun. The scriptural contradiction of the Aristotelian thesis of the eternity of the universe is even more obvious. In regard to the Earth's motion, if one is talking about straight motion downwards, there is no difficulty with Scripture. If one is talking about the Pythagorean or Copernican circular motions, it is less conforming to Scripture, although passages attributing stability to the Earth could be interpreted as attributing perpetuity to it. If one is referring to the Sun and heavens not moving, then the scriptural passages stating the opposite could only be interpreted as accommodating the popular manner of speaking, but such an interpretation should not be adopted "without great necessity."

On November 2, 1612, in a private conversation, Dominican friar Niccolò Lorini attacked Galileo for being inclined to heresy by believing ideas, such as that the Earth moves, which contradict Scripture. However, on November 5, Lorini wrote Galileo a letter of apology. In the fall of 1613, Ulisse Albergotti published a book, *Dialogue... in Which It Is Held... That the Moon Is Intrinsically Luminous...*, containing biblical criticism of Galileo's views.

In December 1613, Christina of Lorraine, the grand duchess dowager and mother of Grand Duke Cosimo II, questioned Galileo's disciple Benedetto Castelli about the compatibility of Galileo's ideas with Scripture. The duchess had been incited by Cosimo Boscaglia, special professor of philosophy at the University of Pisa, who was also present at the meeting. Castelli gave satisfactory answers, but informed Galileo of the incident. Thus, on December 21, 1613, Galileo felt the need to write a long letter to Castelli giving a multifaceted refutation of the scriptural objection to Copernicanism. The details of Galileo's criticism will be discussed later, when we come to the letter which he wrote two years later addressed to the grand duchess Christina herself; in fact, the *Letter to Christina* is an expansion and elaboration of the "Letter to Castelli." But a brief summary will be useful here.

In the "Letter to Castelli," Galileo argued that the scriptural objection to Copernicanism has three fatal flaws. First, it attempts to prove a conclusion (the Earth's rest) on the basis of a premise (the Bible's commitment to the geostatic system) which can only be ascertained with a knowledge of that conclusion in the first place; in other words, the scriptural objection puts the cart before the horse, or in technical terminology, it begs the question. Second, in the biblical objection, the conclusion is derived from the key premise by means of a logically invalid inference, insofar as the Bible is not a scientific authority, and hence its saying something about a natural phenomenon does not make it so, and its statements do not constitute valid reasons for drawing corresponding scientific conclusions; in other words, the objection is a non sequitur, or inferentially invalid. Third, it is questionable whether the Earth's motion really contradicts the Bible, and an analysis of the Joshua passage shows that it cannot be easily interpreted in accordance with the geostatic theory, but accords better with the geokinetic view; in other words, the objection's key premise is false anyway.

Exactly a year after Galileo's "Letter to Castelli," on December 21, 1614, at the Church of Santa Maria Novella in Florence, Dominican friar Tommaso Caccini preached a sermon against mathematicians in general and Galileo in particular; their beliefs and practices allegedly contradicted the Bible and were thus heretical. Caccini illustrated his point by explaining that the biblical passage on the Joshua miracle contradicts the Earth's motion and thus renders belief in it heretical. (Some writers have claimed that Caccini also discussed the suggestive verse in Acts 1:10, "Ye men of Galilee, why stand ye gazing up into heaven?" But there is no contemporaneous evidence that this particular verse was actually mentioned; instead, the story was first told in the eighteenth century.)

Caccini was not the only preacher who attacked Galileo from the pulpit. Another one was Raffaello delle Colombe, also a Dominican friar preaching in Florence, and brother of the philosopher Ludovico delle Colombe, mentioned earlier. Raffaello's sermons were less direct and explicit, insofar as they did not mention Galileo by name; instead they criticized unmistakably Galilean contributions, such as the telescope and sunspots, and they mentioned Galileo implicitly by describing him as "our brilliant Florentine Mathematician." However, Raffaello's sermons were more numerous and persistent, spanning the period 1608–15; and they were more durable, insofar as they were included in a monumental edition of several volumes of sermons on all sorts of topics, edited by Raffaello himself during his own lifetime.[22]

Foscarini's Theological Re-assessment

Not all clergymen held that Copernicanism was incompatible with Scripture, and thus false or heretical. One who did not was the Italian Carmelite friar Paolo Antonio Foscarini (1580–1616), who in 1615 published a book containing a theological defense of Copernicanism. Foscarini was the provincial head of the Carmelites in Calabria and had an ambitious agenda of works in philosophy and theology that tended to be encyclopedic in scope. For the Lent of 1615, he had been invited to preach at the Church of Traspontina in Rome. On his way there from Calabria, he stopped in Naples in January 1615 to supervise the publication of his book. It was written in the form of a letter to the general of the Carmelite order and was entitled *Letter on the Opinion, Held by Pythagoreans and by Copernicus, of the Earth's Motion and Sun's Stability and of the New Pythagorean World System.* As it will emerge, Foscarini's views are similar, although not identical, to those expressed by Galileo in his "Letter to Castelli" and *Letter to the Grand Duchess Christina.*

Foscarini is explicit that his aim is to give a theological defense of the Earth's motion, that is a defense of the geokinetic proposition from the objection that it is contrary to Scripture. He is equally clear that he is in no position to mount the following direct defense: that since Copernicanism is physically true, and since two truths cannot contradict each other, Copernicanism is not contrary to Scripture. This defense is not feasible because the Earth's motion has not been proved with certainty. However, Foscarini is also at pains to repeat frequently the assertion that

Copernicanism is probable or likely true, indeed more probable than the Ptolemaic system, and that this probability is largely the result of Galileo's telescopic discoveries.

To show that Copernicanism is not contrary to Scripture, Foscarini gives several arguments. One of his most important ones is based on the principle of accommodation, which he takes to be uncontroversial and universally accepted: "whenever Sacred Scripture attributes to God or to some creature anything which is otherwise known to be problematic or improper, then it is interpreted and explicated in one of the following four ways";[23] these amount to saying that Scripture is speaking metaphorically or analogically, or is accommodating itself to the common or popular manner of speaking, thinking, perceiving, describing, or believing. Foscarini illustrates this principle with scriptural statements that attribute to God physical attributes such as walking and hands, and emotional states like anger and regret; also with statements that attribute to the Earth ends and foundations; and with those that speak of light and night and day having been created before anything else, of the six "days" of creation, and of the Sun and Moon as the two great luminaries. Moreover, Foscarini is careful to formulate the conclusion of this particular argument by saying that "if the Pythagorean opinion were otherwise true, then it could easily be reconciled with the passages of Sacred Scripture that appear contrary to it... by saying that there Scripture speaks in accordance with our manner of understanding, with the appearances, and with our point of view."[24]

Indeed, such a conditional and relatively weak conclusion is all that follows from the principle of accommodation as stated by Foscarini, which is contingent on a scriptural attribution that is otherwise known to be literally incorrect. Thus, by means of the principle of accommodation, Foscarini does not show, and does not pretend to show, that Copernicanism is indeed compatible with Scripture, but only that if we knew that Copernicanism were true then we could unproblematically reinterpret geostatic statements in Scripture.

Another key argument is based on the principle of limited scriptural authority. Paraphrasing various scriptural passages, Foscarini claims that

> Sacred Scripture...does not instruct men in the truth of the secrets of nature...because [God] has already allowed and decided that the world be occupied with disputations, quarrels, and controversies and be subject to uncertainty in everything (as stated in Ecclesiastes), and that the answer will only come at the end...Thus, its intention is now only to teach us the true road to eternal life.[25]

Here, the conclusion he reaches is that "so consequently we see how and why from the passages already mentioned we cannot derive any certain resolutions in such subjects, and how with this principle we can easily avoid the hits from the first and second group of passages and from any other allegation derived from Sacred Scripture against the Pythagorean and Copernican opinion."[26]

This argument seems a more direct line of reasoning in support of his claim that Copernicanism does not contradict Scripture. For Foscarini is saying that, since Scripture is not an authority on the secrets of nature, scriptural allegations about the Earth's rest and Sun's motion do not entitle us to infer that the Earth is motionless and the Sun moves; thus, we are in no position to assert the Earth's rest on scriptural grounds, and hence the conflict with the Copernican opinion evaporates. In other words, the Earth's motion is not contrary to Scripture because Scripture is not a philosophical (or scientific) authority, and so scriptural assertions that the Earth is motionless do not entail that the Earth is really motionless.

A third argument involves what Foscarini calls the principle of "extrinsic denomination" and the passage in Joshua 10:12–13. This, recall, is the passage where Joshua prays to God to stop the Sun and prevent it from setting, so that the Israelites can have more time of daylight to finish winning a battle against the Amorites; God did the miracle and the Sun stood still for a whole day. The principle states that "many times one says commonly and most properly that a motionless agent moves not because it really moves but by *extrinsic denomination*, namely because with the motion of the subject that receives its influence and action, what also moves is some property which the agent causes in the subject."[27] Applied to the Joshua miracle, we get the following analysis: if the Earth moves and the Sun stands still, sunlight would still move over the Earth's surface, and so it would be proper to say, by extrinsic denomination, that the cause of this moving sunlight itself moves. The Earth's motion can thus be reconciled with the Joshua passage.

Most of the rest of Foscarini's *Letter* consists of arguments attempting to show that various specific scriptural passages that have been alleged to be contrary to Copernicanism can be reconciled with it for various reasons and in various ways.

Foscarini's *Letter* attracted the attention of the Inquisition. By March 1615, the Inquisition had ordered an evaluation of it, and a consultant had written a very critical opinion. Foscarini must have learned something about this censure, and so he wrote a defense of his *Letter* and sent both to Cardinal Robert Bellarmine. On April 12, Bellarmine replied with a gracious but firmly critical letter, which explicitly named Galileo as holding the same position as Foscarini and thus as being liable to the same criticism. Soon thereafter, Foscarini left Rome and returned home with the intention of revising his *Letter* to take such criticism into account. This revision never materialized, because the Inquisition proceedings against Galileo had already started, and they climaxed on March 5, 1616 with an Index decree that condemned and completely prohibited Foscarini's book. Foscarini died on June 10, 1616 "perhaps from a heartbreak," according to one scholar's speculation.[28]

Galileo's Criticism of the Biblical Objection

Although encouraged by Foscarini's book, Galileo was also increasingly concerned with the attacks against his views, especially with the scriptural objections, and especially with the criticism emanating from the pulpit. Moreover, he received a copy of Bellarmine's letter to Foscarini, and from this and other sources he got an inkling that he and Foscarini were being investigated by the Inquisition. So Galileo decided to expand his "Letter to Castelli"; he wrote his essay in the form of a letter to the grand duchess Christina.

The *Letter to Christina* consists of a brief introductory part explaining its origin and purpose; a long central part that takes up in turn a number of distinct questions about the relationship between scriptural interpretation and scientific investigation; and a brief final part in which Galileo engages in some scriptural exegesis meant to show that the Earth's motion is not contrary to Scripture.

In the introductory part, we are told that the letter originated from some unprovoked attacks against Galileo charging him with heresy because he believed in the Earth's motion, and that in it he plans to defend himself from this accusation. It is important to stress the apologetic and defensive character of the letter. Galileo states:

> Now, in matters of religion and of reputation I have the greatest regard for how common people judge and view me; so, because of the false aspersions my enemies so unjustly try to cast upon me, I have thought it necessary to justify myself by discussing the details of what they produce to detest and to abolish this opinion, in short, to declare it not just false but heretical.[29]

The apologia takes the form of the criticism of what we may call the scriptural argument against Copernicanism, and he concludes this part of the letter with the following clear and incisive statement of the objection:

> So the reason they advance to condemn the opinion of the earth's mobility and sun's stability is this: since in many places in the Holy Scripture one reads that the sun moves and the earth stands still, and since Scripture can never lie or err, it follows as a necessary consequence that the opinion of those who want to assert the sun to be motionless and the earth moving is erroneous and damnable.[30]

In the central part of the letter, Galileo addresses himself to the major premise of this argument, that Scripture cannot err. He objects that this proposition is true but irrelevant, because what is relevant is the interpretation of what Scripture says, and scriptural interpretations can indeed err. So the question becomes that of what kind of interpretation, or whose interpretation, if any, is correct. In the various sections of the letter's central part Galileo takes up, in turn, literal interpretation, the interpretation by professional theologians, the interpretation in accordance with the principle of scriptural consensus, the unanimous opinion of Church Fathers, and the official interpretation of the Church (from a pronouncement of the pope speaking *ex cathedra* or from a decision reached by an ecumenical council). A main conclusion here is that scriptural interpretations often presuppose philosophical or scientific claims. Moreover, Galileo distinguishes between questions of faith and morals and questions about the physical universe; he points out that, although Scripture cannot err about the former, when we come to physical questions, it is not so much false as improper to say that Scripture cannot err; the reason is that it is not meant to provide scientific information, and hence it would be equally improper to say that Scripture can be wrong. A central thesis here is that Scripture is not a scientific (or philosophical) authority.

In other words, in this central part of the letter, Galileo interprets the scriptural argument against Copernicanism as essentially an argument from authority to the effect that it is erroneous to believe in the Earth's motion because Scripture says so. He objects that Scripture is not a scientific authority, and therefore even if Scripture does endorse the geostatic thesis, it does not follow that it is true and the geokinetic thesis is false; that is, the reason given for the conclusion is inadequate, even if it were true. He also objects that generally speaking, to know what Scripture really says about physical questions, one has to know the scientific truth about them; this means that to know whether this reason is true, we would have to know whether the conclusion is true, or, as we might say, the argument ultimately begs the question.

The brief final part of the letter may be interpreted as a criticism of truth of the minor premise of the scriptural argument. Galileo tries to show that it is questionable whether Scripture says that the Earth stands still and the Sun moves. He does this by an analysis of several passages that were typically given to support the contrariety thesis. The Joshua miracle (Joshua 10:12–13) is discussed at great length.

Galileo argues that the Joshua passage contradicts the *geostatic* system, whereas it could be given a literal interpretation from the Copernican viewpoint. The passage says that, in response to Joshua's prayer to prolong daylight, God ordered the Sun to stop, and the Sun stood still for a whole day, needed by the Israelites to defeat the Amorites. Galileo points out that in any system, to lengthen the day the diurnal motion must be stopped. Unfortunately, in the geostatic system the diurnal motion belongs not to the Sun, but to the outermost sphere in the universe, either the celestial sphere or the hypothesized sphere called the *primum mobile*. The proper motion that belongs to the Sun is the annual motion, which, being opposite in direction to the diurnal motion, would shorten the day if stopped, making the Sun set that much sooner. It follows that if we take the Scripture literally, the miracle is physically impossible in the geostatic system, whereas if God did the miracle, he should have ordered the *primum mobile* to stop.

By contrast, Galileo argues that in the geokinetic system the miracle could have happened as follows. First, he refers to his own discovery that the Sun is not completely motionless but rotates on its axis with a period of about a month; thus, it makes sense, to begin with, to stop the Sun from moving. To this Galileo adds the speculation that solar rotation probably causes the planetary revolutions, one of which is the Earth's own annual orbital motion; and further that this terrestrial orbital motion is probably connected with the Earth's axial rotation. All of this makes some sense because all these motions are in the same direction in the heliocentric system. Thus, by stopping the Sun's rotation, God could have stopped the Earth's diurnal motion and thus lengthened Joshua's day.

Before we move on to more details of the central part of the *Letter to Christina*, let me summarize my view of its overall conceptual structure (which also corresponds to the basic structure of the "Letter to Castelli," highlighted earlier). This structure amounts to a threefold criticism of the argument that Copernicanism is wrong because Scripture says so: first, Scripture's saying so would not make it so; second, to know what Scripture really says about the physical universe one normally has to know what is physically true; and third, it is questionable whether Scripture does in fact say so.

Galileo begins the central argument of the letter by elaborating several uncontroversial points. The first is that the literal interpretation of Scripture is not always correct since, for example, some scriptural statements about God state or imply that He has eyes, ears, and so on, and we know that it is not literally true. The second point is that the literal interpretation of Scripture is incorrect when it conflicts with physical truths that have been conclusively proved. The third is an explanation for this priority of proved scientific truths over literal scriptural meaning, and it was also universally accepted; the explanation is that, whereas Scripture is the Word of God, which was meant "to teach us how one goes to heaven and not how heaven goes,"[31] the physical universe and the human senses and mind are the *Work* of God, and hence one cannot doubt the truth of physical conclusions grounded on sense experience and conclusive arguments.

From these three points, Galileo thinks it plausibly follows that the literal interpretation of Scripture is not binding when we are dealing with physical propositions that are *capable* of being conclusively proved (even if not proved yet); this consequence follows because doing so would be the more prudent policy, and because what we know is a minute part of what we do not know. Galileo's own words make clear the tentativeness and prudential character of his conclusion: "I should think it would be very prudent not to allow anyone to commit and in a way oblige scriptural passages to have to maintain the truth of any physical conclusions whose contrary could ever be proved to us by the senses or demonstrative and necessary reasons."[32]

Galileo next undertakes an explicit criticism of *theological* authority. He argues that theology is *not* the queen of the sciences because its principles do not provide the logical foundations of the knowledge formulated in other sciences, the way that, for example, geometry does for surveying. Moreover, theologians cannot dictate physical conclusions from the above (i.e., without themselves actually getting involved in physical investigations), any more than a king who is not a physician can prescribe cures for the sick. Nor can theologians tell scientists to undo their own observations and proofs because this is an inherently impossible or self-defeating task. Rather, theologians can and should follow two courses. The first corresponds to already established practice: apropos of conclusively established physical truths, they should strive to show that they are not contrary to Scripture by an appropriate interpretation of the latter. The second would be a rule of interdisciplinary communication. Theologians should presume scientific ideas that are not conclusively proved but contrary to Scripture to be false, and accordingly should try to give a scientific disproof of them; this is desirable because the inadequacies of an idea can be discovered more easily by those who reject it. This ingenious but plausible rule is this section's main methodological conclusion.

Next, Galileo questions the traditional principle that used scriptural consensus combined with the unanimity of the Church Fathers to require acceptance of the literal meaning of physical statements. In other words, he criticizes what earlier I called the anti-Copernican objection from the consensus of Church Fathers. Once again, he makes his fundamental distinction between physical propositions that are and those that are not *capable* of conclusive proof. For the latter the principle makes sense, but for the former the previous considerations suggest that it is not sound. Two new points emerge in this discussion. First, scriptural consensus is not a sign that physical statements are meant to be taken as literally and descriptively true, but rather it is the result of Scripture's desire for consistency, its appeal to common people, and the need to reflect the opinions of the time. Second, the unanimity of Church Fathers is not binding unless it is explicit, unless it is the result of reasoned discussion, and unless it refers to matters of faith and morals.

Finally, the authority of the Church herself comes under discussion. Galileo admits that she does have the power to condemn an idea as heretical, but he notes that "it is not always useful to do all that one can do."[33] Moreover, to make ideas heretical is not the same as making them false; indeed, "no creature has the power of making them be true or false, contrary to what they happen to be by nature and de facto."[34] At any rate, the Church should not be hasty in her condemnation; he hopes that she is not "about to make rash decisions."[35] Before condemning a physical idea, she should examine all the evidence and listen to all the arguments on both sides of the issue, and she should rigorously prove that her interpretation of the relevant scriptural passages is correct. For example, such a rigorous proof should use all the cautious advice elaborated by St. Augustine (quoted below). To avoid potential embarrassment, it might be best to wait until the physical idea is conclusively refuted before declaring it heretical.

One of the most striking features of this central part of the letter is the negative tone of its component conclusions: that the literal interpretation of Scripture is *not* binding in scientific investigation; that theology is *not* the queen of the sciences; that scriptural consensus is *not* a sufficient condition for a literal interpretation; that the unanimity of Church Fathers is *not* necessarily decisive in physical questions; and that the authority of the Church should *not* be hastily applied. This negativity corresponds to the apologetic and critical purpose of the letter, and the general suggestion is, as mentioned earlier, a denial of the scientific (or philosophical) authority of Scripture. But there is an underlying positive idea: the principle of autonomy, according to which scientific investigation can and should proceed independently of Scripture. And from the point of view of the enterprise of understanding Scripture, we get another constructive idea underlying these negative conclusions: that scriptural interpretation often depends on the results of scientific investigation.

A second striking theme is that of prudence and caution, which he adopts from St. Augustine and elaborates further. Galileo's explicit admonitions are, of course, against haste in condemning Copernicanism. But it would also extend to the question of accepting the theory or judging the conclusiveness of its supporting arguments.

Equally striking is the theme involving the distinction between physical propositions that are and those that are not capable or susceptible of conclusive proof. This is obviously the main epistemological distinction, rather than that between propositions that have and those that have not been conclusively proved already. The central issue concerns the former distinction, and Galileo tries to resolve it by arguing that no physical proposition capable of conclusive proof should ever be condemned. The priority of established scientific knowledge that has already been conclusively proved over scriptural statements is a non-issue. From the viewpoint of this uncontroversial principle, there would have been no reason for him to write an essay on the methodology of scriptural interpretation and scientific investigation; rather, the only thing to do would have been to produce or search for the conclusive demonstration. The very fact that he writes this methodological essay indicates that he wants to advocate a (relatively) novel principle.

Besides the argumentative content of this letter and the very fact of writing it, there is a third indication of Galileo's stress on potential demonstrability, as distinct from accomplished demonstration. In 1636, the *Letter to Christina* was published for the first time by some foreign friends but with his cooperation; the edition contained both Galileo's original Italian text and a Latin translation by one of the editors. Now, the stress on demonstrability was explicitly incorporated in the long Latin title of the book, which can be translated as *New and Old Doctrine of the Most Holy Fathers and Esteemed Theologians on Preventing the Reckless Use of the Testimony of the Sacred Scripture in Purely Natural Conclusions That Can Be Established by Sense Experience and Necessary Demonstrations.* Here the crucial phrase is "that can be established," which obviously is not equivalent to "that have been established."

Let me finish with two further considerations on Galileo's *Letter to Christina.* The first involves an aspect of the letter which so far I have largely ignored: the letter is full of references to and quotations from the patristic and theological tradition, such as St. Jerome, St. Thomas Aquinas, and especially St. Augustine. This aspect of the letter could be reconstructed as an argument from authority, or a series of such arguments. This is important for two reasons. Galileo was aware that regardless of how cogent his methodological argument was, his main conclusion (denying scriptural authority for *demonstrable* physical claims) could be taken to be so radical that its novelty needed to be toned down by trying to root it in tradition. Thus, in the just quoted title of the 1636 edition of the letter, the initial part of the title ("*New and Old,*" in Latin a single word "*Nov-antiqua*") stresses precisely the two-fold aspect of being partly radical and partly traditional. And secondly, the passages quoted from Augustine are so crucial that they played a significant role in the subsequent history of scriptural hermeneutics. For example, in 1893, Pope Leo XIII's encyclical *Providentissimus Deus,* without even mentioning Galileo, put forth a Galilean view of the role of Scripture in scientific investigation, and appealed to the same Augustinian passages as had been quoted by Galileo.

Two passages serve to give a flavor of Galileo's appeal to St. Augustine. One is Augustine's version of the principle of nonscientific authority of Scripture: "it should be said that our authors did know the truth about the shape of heaven, but that the Spirit of God, which was speaking through them, did not want to teach men these things which are of no use to salvation."[36] The other is Augustine's versions of the principle of the priority of demonstrated physical truth: "whenever the experts of this world can truly demonstrate something about natural phenomena, we should show it not to be contrary to our Scriptures."[37]

The second observation is that in the *Letter to Christina* there are several passages that undeniably appear to be inconsistent with the principle that Scripture is not an authority in astronomy or natural philosophy. Some scholars stress this inconsistency and end up attributing to Galileo an incoherent position, which includes a principle of priority of Scripture. But in my view the alleged Galilean inconsistencies are more apparent than real, and Galileo's alleged incoherence reflects the inadequate and insufficiently deep analysis of such scholarly accounts.

If such Galilean assertions were expressions of the principle of the general priority of Scripture, they would seriously undermine the apologetic purpose of the letter, which was after all to refute the scriptural objection to Copernicanism by arguing (among other criticisms) that the objection is a non sequitur because scriptural statements about the Earth's or Sun's motion or rest do not entail that the Earth or Sun really moves or rests. However, if these passages were regarded essentially as attempts to define more precisely the proper scope of the principle of non-scientific authority of Scripture, then such an interpretation would conform with the apologetic purpose of the letter, and so would be preferable to the alternatives that undermine that purpose. I believe such an interpretation would be along the following lines.[38]

The first of the problematic passages declares that scriptural assertions have priority over other assertions in regard to historical questions. The second passage states that Scripture has priority over *unprovable* assertions in regard to physical and natural phenomena. The third claims that, for theologians, scriptural assertions have priority over *unsupported* assertions in all other writings. In other words, Scripture is a superior authority regarding (1) historical questions that depend on balancing probabilities of testimony; (2) undecidable questions about physical reality; and (3) unsupported assertions on any topic in any book. So although Galileo denies the scientific (astronomical, or philosophical) authority of Scripture, he accepts its authority not only for questions of faith and morals, but also for the weighing of probable testimony in history, for undecidable questions in natural philosophy, and for questions of presumption of truth for unsupported claims. These are important nuances, complications, and qualifications in Galileo's position, but none of this undermines his criticism of the scriptural objection to Copernicanism or his principle of scriptural irrelevance in astronomical research.

CHAPTER 5

THE EARLIER INQUISITION PROCEEDINGS (1615–1616)

Clerical Actions and Galilean Responses

In February 1615, the Dominican friar Niccolò Lorini filed a written complaint against Galileo with the Inquisition in Rome, enclosing his letter to Castelli as incriminating evidence. Lorini charged that Galileo's letter to Castelli was theologically questionable in several ways, but chiefly in claiming that Scripture has no role in scientific investigation, and in defending Copernicus's view that the Earth moves (which was contrary to Scripture). Then in March of the same year, another Dominican, Tommaso Caccini, who had attacked Galileo from the pulpit in December 1614, made a personal appearance before the Roman Inquisition. In his deposition, he charged Galileo with suspicion of heresy, based not only on the content of the letter to Castelli, but also on the *Sunspots* book (1613); and he mentioned some hearsay evidence, both general and specific, involving two individuals named Ferdinando Ximenes and Giannozzo Attavanti. The Roman Inquisition responded by ordering an examination of these two individuals and of the two mentioned writings.

In the meantime, Galileo was writing for advice and support to many friends and patrons who were either clergymen or had clerical connections. He had no way of knowing the details of the Inquisition proceedings, which were a relatively well-kept secret. However, he happened to learn about Lorini's complaint, and he knew about Caccini's original sermon, which of course had been public.

Moreover, as we have seen to some extent, Galileo wrote and started to circulate privately three long essays on the issues. One, the *Letter to Christina*, dealt with the religious objections and was an elaboration of the letter to Castelli, which was thus expanded from 8 to 40 pages. Another, now known as "Galileo's Considerations on the Copernican Opinion," began to sketch a way of answering the epistemological and philosophical objections, which Galileo had not done before; the importance of such objections had recently been stressed in the letter by Cardinal Robert Bellarmine, addressed to Foscarini, but also indirectly meant for Galileo. And the third essay, the "Discourse on the Tides," was an elementary discussion of the scientific issues, in the form of a new physical argument in support of the Earth's motion, based on its alleged ability to explain the existence of the tides and of the trade winds; this essay was an anticipation of what Galileo would later elaborate in the Fourth Day of the *Dialogue*.

Furthermore, in December 1615, after a long delay due to illness, Galileo went to Rome of his own initiative, to try to clear his name and prevent the condemnation of Copernicanism. The results were mixed at best. On the one hand, at a personal and informal level, he was generally well received by friends, acquaintances, and various powerful clergymen. On the other hand, at an official and formal level, the machinery of the Inquisition had been put in motion and was following its own procedures, practices, and rules. But before we come to the Inquisition actions and decisions, it is worth noting one feature of Galileo's own activities and discussions.

A memorable description of Galileo's behavior is found in a letter to Cardinal Alessandro D'Este by his secretary Antonio Querenghi. The cardinal was then residing in his hometown of Modena, but he employed a secretary who lived in Rome and was regularly writing reports about what was happening in the eternal city. In a letter dated January 20, 1616, Querenghi included this passage:

> Your Most Illustrious Lordship would really like Galileo if you heard him argue, as he often does, surrounded by fifteen or twenty people who launch cruel attacks against him, now about one thing and now about another. But he is so well fortified that he is amused by them all; and although he does not persuade them, on account of the novelty of his opinion, nevertheless he proves the invalidity of most arguments with which his enemies try to bring

> him down. On Monday in particular, at the house of Mr. Federigo Ghisilieri, his arguments were astonishing; what I liked most was that, before answering the contrary reasons, he amplified and strengthened them with new grounds of great plausibility, so that after he destroyed those reasons, his opponents would appear more ridiculous.[1]

This Galilean behavior has been widely interpreted as an illustration of the sophist's art of winning an argument by first defending one thesis, then defending the opposite, and thus confusing and ridiculing opponents. But the truth is almost the opposite. If we look carefully at what Querenghi is saying in this letter, he is clear that Galileo was not being too successful at convincing his interlocutors to accept the Copernican hypothesis of the Earth's motion. It is equally clear that Galileo was convincing his listeners that the objections to the Earth's motion were invalid or ineffective. So he was succeeding in refuting the objections against the Earth's motion, but his success at refutation was not total and did not apply to all objections; he could only refute "most" objections. Yet Querenghi notes that Galileo followed a very striking method of refutation: before responding to the anti-Copernican objections, he strengthened them. In his defense of Copernicanism, Galileo was operating on three levels: he was trying to provide positive or constructive evidence or reasons supporting the Earth's motion; he was trying to articulate answers to or criticisms of the objections or counter-arguments; and he was trying to formulate the opposite arguments as strongly and plausibly as possible before refuting them.

This is a very sophisticated, powerful, and admirable method of argument, very different from the way Galileo has often been depicted, as ridiculing opponents by means of the trick of first persuading them of one thing and then proving to them the opposite. On the contrary, the Galilean technique of strengthening objections before refuting them shows that the objections, although invalid, are serious, important, and plausible, and therefore that the opponents who believe the contrary thesis are reasonable people. Far from ridiculing opponents, this is a way of ennobling them, paying them respect, and enhancing their standing and credibility. Galileo's technique is, in fact, the antithesis of the widespread (then and now) practice of demonizing one's opponents; it is the antidote to the straw-man fallacy.

Later, we shall see that Galileo practiced this technique in the *Dialogue* and formulated it into a principle in one of his 1633 depositions, and that it represents a universal and perennial lesson from the Galileo affair. For now, let us go back to the earlier Inquisition proceedings and their results.

The consultant who examined the letter to Castelli reported that in its essence its hermeneutical views did not deviate from Catholic doctrine.[2] The cross-examination of the two witnesses, Ximenes and Attavanti, exonerated Galileo from the hearsay evidence; his utterance of heresies was found to be baseless. And the examination of his work on *Sunspots* failed to reveal any explicit assertion of the Earth's motion or other presumably heretical assertion, if indeed the Inquisition officials examined this book. However, in the process, the status of Copernicanism had become enough of a problem that the Inquisition felt it necessary to consult its experts for a formal opinion.

On February 24, 1616, a committee of eleven consultants filed their report. Their unanimous opinion was two-fold: from a scientific or philosophical point of view, Copernicanism was false; and from a theological point of view, it was heretical or erroneous since it contradicted Scripture. In a way, much of the tragedy of the Galileo affair stems from this opinion, which even Catholic apologists seldom if ever defend nowadays.

Although indefensible, if one wants to understand how this opinion came about, one must recall all the traditional arguments against the Earth's motion, based on astronomical, empirical, physical, mechanical, philosophical, and epistemological considerations, as well as the theological and scriptural ones. Moreover, one must view the judgment of heresy in the light of the two objections based on the words of the Bible and on the consensus of the Church Fathers; in the light of the traditional hierarchy of disciplines, which made theology the queen of the sciences, and which had been reaffirmed at the Fifth Lateran Council in 1513; and in the light of the Catholic Counter-Reformation rejection of new and individualistic interpretations of the Bible.

However, the Inquisition must have had some misgivings about this assessment by the committee of eleven consultants. It did not accept the heresy part of their recommendation, and so issued no formal condemnation that the Copernican doctrine was a heresy. Instead some milder consequences followed, some pertaining to Galileo personally, and some pertaining to the status of the Copernican doctrine.

Pope Paul V's Orders

On February 25, 1616, there was a meeting of the Inquisition presided over by Pope Paul V. The decision on the Galileo case was communicated to the Inquisition's commissary and the assessor by the cardinal secretary. This communication is recorded in a document included in the Vatican dossier of Galilean proceedings.[3]

These papal orders envisaged three possible steps. First, Cardinal-Inquisitor Bellarmine (Figure 13) would give Galileo an informal and friendly warning to abandon the heliocentric geokinetic doctrine. The content of this warning is expressed in terms of the notion of abandonment, which means to stop believing, accepting, or holding this doctrine, and so refers to an internal mental state.

The second step would be taken if Galileo rejected Bellarmine's warning. Then the Inquisition's commissary was supposed to intervene and issue Galileo an injunction or precept. This would be a formal and official step meant to hold Galileo legally responsible and liable for his external public cognitive activities. The content of this injunction would be much more stringent than the warning: Galileo would be completely prohibited to teach, defend, or even discuss the doctrine. Teaching refers to two relatively distinct activities: one is the explanation of a doctrine for the sake of conveying an understanding of it; the other is a discussion of the doctrine to render it plausible or acceptable, which would be essentially equivalent to supporting it. Similarly, defending can refer to two things: primarily, answering or refuting objections or counter-arguments; but secondarily,

Figure 13. Cardinal Robert Bellarmine (1542–1621)

supporting a doctrine with evidence or reasons. Thus, the content of the pope's intended injunction was meant to prohibit Galileo from discussing Copernicanism, understanding that a discussion can include explanatory teaching, supporting with reasons, and defense from objections, among other activities.

Note, however, that discussion does not encompass belief or acceptance. One can discuss a doctrine without necessarily accepting or rejecting it, and conversely one can accept or reject a doctrine, and yet refrain from discussing it. Accordingly, the intended injunction does not include the abandonment of the opinion, but rather refers to external and public activities, distinct from internal mental states.

The third step mentioned in the document refers to what would happen in case Galileo rejected the commissary's injunction, namely imprisonment. It's unclear how literally one should take the notion of imprisonment, if that meant indefinite prison without a trial. But it is meant to convey at least the notion of being arrested.

In short, on February 25, Pope Paul ordered Cardinal Bellarmine to give Galileo a friendly warning to stop believing in the geokinetic doctrine; this was an order regarding his private mental state. If, and only if, Galileo rejected this warning, the Inquisition's commissary was to give him a formal injunction to completely refrain from discussing the doctrine; this was an order regarding his public cognitive activities. And if, and only if, Galileo rejected this injunction was he to be arrested.

Commissary Seghizzi's Injunction

The following day, February 26, these papal orders were carried out. One description of what happened is given in a memorandum written by a notary and filed among the documents in the Vatican dossier of Galilean trial proceedings.[4]

According to this document, Galileo was called to Bellarmine's residence. Here, Bellarmine warned Galileo that he should abandon the heliocentric doctrine, and, immediately thereafter, the Inquisition's commissary Michelangelo Seghizzi enjoined Galileo to abandon completely the doctrine and henceforth not to hold, teach, or defend it in any way whatever. It is also reported that Galileo acquiesced after receiving the commissary's injunction.

The most striking thing about this document is that it reports the occurrence of the first two steps of the papal orders, without motivating the second step. Bellarmine first issued his friendly warning, but we are told nothing about a rejection on Galileo's part, and yet, allegedly, Commissary Seghizzi immediately went on to issue him a formal injunction, and we are told that Galileo accepted the latter. Of course, Seghizzi's injunction presupposes that Galileo rejected Bellarmine's warning; but if that happened then this memorandum should have reported that development. The fact that the memorandum does not report it suggests that it did not happen—that Galileo accepted Bellarmine's warning; but then Commissary Seghizzi's injunction would be a deviation from the papal orders, and thus an arbitrary or illegal action.

However, this is not its only irregularity. Another question concerns the content of this injunction. Seghizzi ordered Galileo, among other things, to completely abandon and not to hold in any way the Copernican opinion. But the pope's intended injunction said nothing about abandonment. The papal orders did mention abandonment as the main point of Bellarmine's warning, but not as part of the commissary's potential injunction. Moreover, with regard to holding, insofar as it means believing, the pope had intended its prohibition to be also part of Bellarmine's warning. Admittedly, insofar as holding also means supporting, the pope did intend to include it in the injunction, being encompassed by some of the connotations of teaching and defending; and these notions are included in Seghizzi's injunction, and so there is no problem in that regard. In short, Seghizzi's precept is in one sense more stringent than the pope's intended injunction, insofar as the former prohibits a private mental state, which the latter had not planned to be the subject of a formal injunction, but rather of a previous friendly warning.

To make this situation even more problematic, it also turns out that Seghizzi's injunction is in another sense *less* stringent than the pope's intended precept. Pope Paul had envisaged that if Galileo rejected Bellarmine's friendly warning to stop accepting Copernicanism, then Galileo would be issued the formal injunction to refrain from discussing it. The key point now is that a prohibition of discussion clearly includes a prohibition of refutation and criticism, but the latter are *not* excluded by Seghizzi's injunction. If Galileo had wanted to publish a technical discussion of why Copernicanism was false, or otherwise flawed, this would not be a way of teaching, supporting, or defending Copernicanism, and so would be allowed by Seghizzi's injunction; but it would still be a discussion, and hence banned by the pope's intended injunction.

And there is one further irregularity in Seghizzi's precept compared to that intended by the pope. This involves the consequences of Galileo's rejection of the injunction. For the pope, the stated threat was imprisonment, which perhaps could be taken to mean arrest; but in the February 26 document the stated threat is prosecution by the Inquisition. And the latter is certainly weaker than the former.

Given these discrepancies with papal intention,[5] the legitimacy of Commissary Seghizzi's precept is certainly questionable. These irregularities also cast doubt on the factual accuracy of the document regarding what really happened at Bellarmine's house. Such legal impropriety and/or factual inaccuracy remain even if the document is authentic and not a forgery. Whether the document is genuine is indeed an important topic, but need not be examined here. I am inclined to think that it is authentic. But the question of documentary authenticity should not be confused with the question of legal validity or with the question of factual accuracy.

The Index's Decree

The legality and/or accuracy of Seghizzi's injunction is further undermined by two things which Cardinal Bellarmine said and did after meeting with

Galileo and Seghizzi on February 26. They are known to us from two crucial documents in which Bellarmine explicitly addresses the question of what happened that day. However, these documents also mention a public decree published by the Congregation of the Index. This is an important document in its own right, and reflects one of the two main decisions reached by the Inquisition in 1616.

On March 5, 1616, the Index issued a decree which did not mention Galileo at all, but contained various censures of the geokinetic doctrine and prohibitions of Copernican books.[6] Although dealing mostly with books, the decree also contains a doctrinal pronouncement. The doctrine of the Earth's motion is declared scientifically false and theologically contrary to Scripture. Note that the latter theological censure is not equivalent to heresy, but is a weaker form of condemnation. Moreover, the book published by Foscarini in 1615 is condemned and permanently banned, because it argued that the Earth's motion is true and compatible with Scripture; this book was trying to do something that goes explicitly against both parts of the Index's doctrinal declaration. However, Copernicus's own book is merely banned temporarily until appropriately revised; this decree does not explain why.

The revisions were explained in a subsequent decree by the Index, published four years later.[7] The gist of the 1620 decree was that Copernicus's book was valuable from the viewpoint of astronomical calculation and prediction; that the book was treating the Earth's motion primarily as a hypothetical construct, and not as a description of physical reality; that one could easily delete or rephrase the few passages where the book treated the Earth's motion as physically real or compatible with Scripture; and that specific instructions could be given about how to delete or rephrase the dozen passages where this happened. Such clarifications and corrections may be taken as being implicit already in the 1616 decree.

Thus, we can now better understand a final point concerning books in general, explicitly made in the 1616 decree. It states that all other books teaching the same doctrines are similarly prohibited. In other words, books like Foscarini's were also condemned and permanently banned, whereas books like Copernicus's were suspended until and unless revised, to render them hypothetical. This meant presumably that no Catholic could do what Foscarini had tried to do, namely to support the physical reality of the Earth's motion, or to defend it from the objection that it contradicts Scripture. On the other hand, Catholics were free to treat, and support and defend, the Earth's motion as a hypothesis for the convenience of saving the appearances and making mathematical calculations and astronomical predictions, as Copernicus had allegedly done.

Cardinal Bellarmine's Warning

Let us now go back to the two pieces of evidence from Bellarmine. On March 3, 1616, at an Inquisition meeting presided over by the pope, Bellarmine reported what he had done to carry out the pope's orders of the previous week.[8] He says that Galileo accepted his warning to abandon the geokinetic doctrine. He says nothing about any injunction issued to Galileo by Commissary Seghizzi; and of course, this is in accordance with the pope's earlier orders, which made the injunction contingent on Galileo's *rejection* of the warning. The minutes of this meeting also report that the pope ordered the publication of the Index's decree, which is summarized in terms of the distinction between the suspension of books like Copernicus's and the prohibition of books like Foscarini's.

The other action by Bellarmine that sheds light on what happened at his meeting with Galileo and Seghizzi involves a certificate, dated May 26, 1616.[9] Bellarmine wrote it at the request of Galileo, who in April received letters from friends saying that there were rumors to the effect that he had been put on trial, condemned, forced to recant, and given appropriate penalties by the Inquisition.

In the document, Bellarmine speaks only of his warning, and not of a formal injunction by Commissary Seghizzi. However, here Bellarmine elaborates the content of his own warning. Previously, he had described it as ordering Galileo to abandon Copernicanism, meaning to stop believing in it. This certificate contains a clarification involving a reference to the Index decree of March 5, 1616, and an explicit mention of the notions of defending and holding.

The meaning of these notions of defending and holding here may be taken to be the same as the meaning we took them to have earlier; they refer to believing, supporting with reasons, and defending from objections. But the reference to the Index decree specifies their meaning even further. Bellarmine is saying partly that what is prohibited is to believe, support, or defend the Earth's motion as true or as compatible with Scripture, in line with the condemnation and complete prohibition of Foscarini's book. Bellarmine is also saying that there is no prohibition on believing, supporting, or defending the Earth's motion as a convenient hypothesis for saving the appearances and making calculations and predictions. This additional permissive clarification is in accordance with the mere suspension of Copernicus's book, published by the Index two months earlier, and in accordance with the subsequent correction of that book as specified in the 1620 decree.

In short, Bellarmine's certificate of May 26, 1616, gives us a fourth version of an Inquisition order to Galileo. In a sequence of increasing rigor, we have the following. The mildest order is the warning intended in the pope's February 25 decision, to abstain from believing the Earth's motion. The second-weakest order is the one described by Bellarmine in his May 26 certificate: namely, to abstain from believing, supporting, or defending the doctrine as true or as compatible with Scripture. Third, there is the formal injunction supposedly delivered by Commissary Seghizzi according to the February 26 document: to abstain from believing, supporting, defending, or teaching the doctrine in any way whatever. Finally, there is the injunction intended by Pope Paul in case of Galileo's rejection of Bellarmine's warning: to completely abstain from discussing the doctrine.

Publishing a refutation of Copernicanism would violate Paul's intended injunction, but not any of the other three orders. An explanation of the content, the arguments in favor, and the arguments against Copernicanism would violate Seghizzi's injunction, as well as the one intended by the pope, but neither of the other two orders; such an explanation would be a so-called disputation, in the traditional sense. The same would be true if one were to give a demonstration that the geokinetic doctrine is better than the geostatic one as a hypothesis for saving the appearances and making calculations and predictions; it would violate the two stronger orders, but not the two weaker ones. A simple and clear violation of the last three orders, but not of the first, would be a defense of Copernicanism from the objection that it contradicts Scripture; and this of course is found in Foscarini's book and in Galileo's unpublished letters to Castelli and to Christina.

However, the version of Bellarmine's warning embodied in his certificate given to Galileo is not only the most likely to be factually accurate but also the most plausibly legitimate one. For a start, Bellarmine's warning, as formulated in the certificate, although different from the pope's intended warning, may be seen as an elaboration of it. The strengthening of Bellarmine's order to include the external acts of supporting and defending, besides the internal mental state of believing, is meant to explain its meaning and to render it applicable. When, at Bellarmine's house, Galileo was first confronted with the warning to abandon or to stop believing the geokinetic doctrine, he is likely to havc reacted by saying that he did not really believe the doctrine, because he did not think the issue was settled yet or that the relevant arguments were conclusive. What he had done so far was to discover some new evidence in its favor and to defend it from some old objections. At this point, then, Bellarmine specified and added that the decisions by the pope, the Inquisition, and the Index implied that he (Galileo) was supposed to stop supporting and defending the doctrine as true or as compatible with Scripture. But Bellarmine also weakens the warning by telling Galileo that he could support and defend it as a convenient hypothesis, to save the appearances and to make calculations and predictions.

Secondly, as a cardinal-inquisitor, Bellarmine may be taken to have the authority to interpret, explain, and apply the pope's intended warning in this manner. As a cardinal-inquisitor, he was both one of the judges of the tribunal and one of the legislators of the relevant rules. Bellarmine was clearly the most respected and authoritative of these judges and legislators.

Pope Paul V himself reached his decision by discussing the matter with all the cardinal-inquisitors, but primarily with Bellarmine. So he was not abusing or perverting the papal orders, but primarily interpreting and applying them.

Neither of these two points apply to Seghizzi's precept. His deviations from the pope's intended injunction are significant. His addition of a prohibition on Galileo's belief enables the precept to enter the area of mental states, which is arguably off limits to administrative and legal measures.[10] On the other hand, Seghizzi's failure to prohibit discussion of the topic would allow Galileo to do things (such as criticizing Copernicanism) which Pope Paul had intended to stop and exclude. Moreover, in his position as commissary, Seghizzi did not have the authority to modify the orders of his superiors in such significant ways. It's not clear that even a cardinal-inquisitor like Bellarmine could have properly made such modifications to a decision reached at an Inquisition meeting presided over by the pope. The point is that Seghizzi's precept does not merely interpret, apply, or clarify the pope's orders, but perverts and subverts them. Thus, the irregularity of Seghizzi's injunction remains.

In conclusion, the 1616 Inquisition's orders to Galileo regarding Copernicanism were numerous and somewhat confusing. Some of these orders were actually given, and legitimately so; chiefly, Bellarmine's warning not to hold, support, or defend Copernicanism as true or as compatible with Scripture, but only as a hypothesis. Some orders were intended to be given, but were not actually delivered; such is the case for Pope Paul's injunction not to discuss Copernicanism. Some orders were allegedly, but probably not actually, given, and in any case illegitimate if issued; that applies to Commissary Seghizzi's order not to hold, support, or defend Copernicanism in any way. However, this multiplicity and confusion contributed to generating the later proceedings in 1632–3, and will help us understand and explain them.

CHAPTER 6

THE *DIALOGUE ON THE TWO CHIEF WORLD SYSTEMS* (1632)

Resuming the Copernican Discussion

For the next several years, Galileo did refrain from defending or explicitly discussing the geokinetic theory, although he did discuss it implicitly and indirectly in the context of a controversy about the nature of comets in *The Assayer* (1623). Even the publication of the corrections to Copernicus's book in the Index's decree of 1620, which gave one a better idea of what was allowed and what not, did not motivate him to resume the earlier struggle. The death of both Cardinal Bellarmine and Pope Paul V in 1621, and the election of Pope Gregory XV did result in some encouragement, for example when Galileo was consulted about astronomical matters by the cardinal nephew and Vatican secretary of state; but those developments were not enough for a significant change.

The event that put an end to the interlude took place in 1623, when Gregory XV died and Cardinal Maffeo Barberini was elected Pope Urban VIII (see Figure 14). Urban was a well-educated Florentine, and in 1616 he had been instrumental in preventing the direct condemnation of Galileo and the formal condemnation of Copernicanism as a heresy. He was also a great admirer of Galileo, and in 1620 he had even written a poem in praise of Galileo. He now employed as personal secretary one of Galileo's closest acquaintances, Giovanni Ciampoli (1589–1643). Furthermore, at about this time, Galileo's book on the comets, *The Assayer*, was being published in Rome by the Lincean Academy, and so it was decided to dedicate the book

Figure 14. Maffeo Barberini (1568–1644), Pope Urban VIII (1623–44)

to the new pope. Urban appreciated the gesture and liked the book very much. Finally, as soon as circumstances allowed, in the spring of 1624, Galileo went to Rome to pay his respects to the pontiff; he stayed about six weeks and was warmly received by Church officials in general and the pope in particular, who granted him weekly audiences.

The details of the conversations during these six audiences are not known. There is evidence, however, that Urban VIII did not think Copernicanism to be a heresy, or to have been declared a heresy by the Church in 1616. He interpreted the Index's decree to mean that the Earth's motion was a rash or dangerous doctrine whose study and discussion required special care and vigilance. He thought the theory could never be proved to be necessarily true, and as we saw earlier, his favorite argument for this skepticism was the divine-omnipotence objection: God being all powerful, he could have created a world in which the Earth did not move, so asserting that the Earth must move is to wish to limit God's power.[1] This argument, together with his interpretation of the decree of 1616, must have reinforced his liberal inclination that, as long as one exercised the proper care, there was nothing wrong with the hypothetical discussion of Copernicanism, with treating the Earth's motion as a hypothesis and studying its consequences, and its utility for making astronomical calculations and predictions.

At any rate, Galileo must have gotten some such impression during his six conversations with Urban, for upon his return to Florence he began working on a book. This was in part the work on the system of the world which he had conceived at the time of his first telescopic discoveries, but it now acquired a new form and new dimensions in view of all that he had learned and experienced since. His first step was to write and circulate privately a lengthy reply to the anti-Copernican essay written in 1616 by Francesco Ingoli. This Galilean "Reply to Ingoli," as well as his earlier "Discourse on the Tides," were incorporated into the new book. After a number of delays in its writing, licensing, and printing, the work was finally published in Florence in February 1632, with the title *Dialogue on the Two Chief World Systems, Ptolemaic and Copernican.*

The author had done a number of things to avoid trouble, to ensure compliance with the many restrictions under which he was operating, and to satisfy the various censors who issued him permissions to print.

To emphasize the hypothetical character of the discussion, Galileo had originally entitled it "Dialogue on the Tides" and structured it accordingly. It was to begin with a statement of the problem of the cause of tides, and then it would introduce the Earth's motion as a hypothetical cause of the phenomenon; this would lead to the problem of the Earth's motion, and to a discussion of the arguments pro and con, as a way of assessing the merits of this hypothetical explanation of the tides.[2] However, the book censors, interpreting and acting on the pope's wishes, did not like such a focus on the tides, which suggested a realistic interpretation of Copernicanism and the potential reality of the Earth's motion. Instead, they wanted to make the book look like a vindication of the Index decree of 1616. Thus, the book's preface, whose content must be regarded as originating primarily from the pope and the censors and only secondarily from Galileo, claimed that the work was being published to prove to non-Catholics that Catholics knew all the arguments and evidence about the scientific issues, and so their decision to believe in the geostatic theory was motivated by religious reasons and not by scientific ignorance. It went on to add that the scientific arguments seemed to favor the geokinetic theory, but that they were inconclusive, and thus the Earth's motion remained a hypothesis.

Galileo also complied with the explicit request to end the book with a statement of the pope's favorite argument, the objection from divine omnipotence. Such compliance reflected in part Galileo's readiness and willingness to be cooperative and accommodating. It also reflected his judgment and recognition that there was something right about the pope's favorite objection.

Moreover, to make sure he would not be seen as holding or defending the truth of the geokinetic thesis (which he knew he had been forbidden to do), Galileo wrote the book in the form of a dialogue among three speakers: Simplicio, defending the geostatic side; Salviati, taking the Copernican view; and Sagredo, who is an uncommitted layperson who listens to both sides and accepts the arguments that seem to survive critical scrutiny. Again, this aspect of the book also corresponded to Galileo's deeply held views about the nature of knowledge. It reflected his view that knowledge is a process that requires such things as critical reasoning, argumentation, open-mindedness, fair-mindedness, etc. In this regard, Galileo's epistemology overlapped (without, however, being identical) with that of the ancient Greek philosopher Plato, who also wrote all his works in dialogue form.

Likewise, in many places throughout the book, usually at the end of a particular topic, the Copernican Salviati utters the qualification that the purpose of the discussion is information and enlightenment, and not to decide the issue, which is a task to be reserved for the proper authorities. This, too, was partly prudential precaution, and partly logical or methodological judgment. Galileo did not want to be seen as contradicting either the Index's anti-Copernican decree or Bellarmine's personal warning to him, but he also judged that the case against the geostatic thesis and in favor of Copernicanism, however strong and probable, was not conclusive and decisive.

Galileo moreover obtained written permissions to print the book, first from the proper Church officials in Rome (when the plan was to publish the book there), and then from the proper officials in Florence (when a number of external circumstances dictated that the book be printed in the Tuscan capital).

The *Dialogue* is essentially a critical examination of all scientific and philosophical arguments on both sides of the Copernican controversy. The discussion thus includes all the astronomical, mechanical, physical, epistemological, and methodological arguments for and against the Earth's motion. It usually consists of a statement or formulation of the argument, often a quotation from some author whom Galileo intends to criticize; an interpretation or clarification designed to convey a deeper understanding of the argument; an evaluation or assessment of whether the argument is valid or cogent, and of its strengths and weaknesses; and an analysis meant to explain and justify his interpretations and evaluations.

On the other hand, the book's critical examination does *not* include the religious, theological, and biblical objections to the Earth's motion. Galileo had not forgotten that, in light of the earlier Inquisition proceedings, he was not supposed to defend Copernicanism from these objections; and it was also obvious that the more liberal atmosphere resulting from Pope Urban VIII did *not* extend to this topic. The book's ending does mention the divine-omnipotence argument, as required by the pope and the censors; and, accordingly, there is no critical analysis of that argument. (The *Dialogue* also contains one brief passage mentioning the biblical objection, in the context of reporting the arguments advanced by an anti-Copernican author; but Galileo summarily drops any critical discussion of this topic, and instead briefly expresses his reverence for the Bible and his unwillingness to mix science and religion.)

The book is divided into four parts or chapters, called "Days" to reflect the fiction that the three interlocutors met on four different days to focus on four distinct but interrelated problems. The First Day examines the Earth–heaven dichotomy, whose truth would have made the Copernican system impossible. The Second Day considers the diurnal motion, whether it belongs only to the Earth or to all heavenly bodies except the Earth. The Third Day focuses on the annual motion, whether it belongs to the Earth or to the Sun. The Fourth Day discusses the problem of the tides, why they take place and how the Earth's motion can cause them to come about.

First Day: Earth–Heaven Dichotomy

In the First Day, Galileo criticizes the Aristotelian arguments in favor of the Earth–heaven dichotomy and elaborates an argument showing that there is no essential difference between terrestrial and heavenly bodies. Bear in mind that the contradiction between the Aristotelian thesis and Copernicanism was explicit, direct, and palpable. The primary evidence for the similarity between Earth and heaven had been available since the telescopic discoveries, but Galileo had not had the occasion to elaborate the argument explicitly. He argues as follows.

The telescopic observation of the Moon shows that its surface is not perfectly smooth, but rather rough, full of mountains and valleys, made of a substance that is opaque, non-luminous, and casts shadows; this is very much like the Earth. The observation of sunspots points in the same direction: they are not previously unknown planets circling the Sun, but phenomena on the surface of the Sun; they appear as dark spots, grow in size, and then gradually fade away; they occur somewhat irregularly, but while a particular sunspot lasts, it appears to move across the solar disk in such a way as to indicate that the Sun undergoes an axial rotation with a period of about one month. In the First Day, Galileo also briefly mentions some naked-eye observations that did not require a telescope: the novas that became visible in 1572 and in 1604 and that appeared to be instances of generation and decay in the heavens. Moreover, later (in the Third Day) Galileo analyzes at great length the observations of the 1572 nova by a dozen astronomers, to refute the attempt to show that this nova was located below the Moon, in the sublunary region.

Galileo is clear that to establish the similarity between Earth and heaven does not amount to proving the key Copernican thesis that the Earth moves. It remains to be shown whether the similarity between terrestrial and heavenly bodies extends to motion.

Galileo is equally clear that his observational argument for the Earth–heaven similarity merely refutes the dichotomy thesis, thus depriving the Aristotelians of a reason to reject Copernicanism. The argument as such does not even tell us what is wrong with the reasoning which led the Aristotelians to believe in the Earth–heaven dichotomy. However, Galileo thinks that it is equally important to understand how their reasoning went wrong, and about half of the First Day is devoted to such a critical analysis.

For example, one of their arguments was itself an observational argument: the heavenly region is devoid of physical changes (other than regular circular motion) because no such changes have ever been observed in the heavenly bodies. Galileo points out that this argument presupposes that what is observed by the senses normally corresponds to reality. He also points out that Aristotle himself explicitly asserted such a principle on many occasions. Galileo can thus formulate the memorable criticism that "it is more in accordance with Aristotle to philosophize by saying 'the heavens are changeable because so the senses show me' than if you say 'the heavens are unchangeable because theorizing so persuaded Aristotle'."[3]

But Aristotle also had some theoretical arguments in support of the Earth–heaven dichotomy, and they too deserve critical scrutiny. In fact, Galileo begins the First Day with a critical analysis of those theoretical arguments.

At the most fundamental and abstract level, the Earth–heaven dichotomy was based on Aristotle's distinction between two kinds of natural motions. He claimed that there are two distinct kinds of natural motions: the first is straight, toward or away from the center of the universe, and characteristic of elementary terrestrial bodies such as earth, water, air, and fire; the other is circular, around the center of the universe, and characteristic of heavenly bodies made of a fifth substance named aether.

This Aristotelian distinction between straight and circular natural motions was the key element of the Earth–heaven dichotomy, insofar as the distinction *per se* divided the universe into two regions where bodies behaved in accordance with radically different laws. Moreover, the distinction helped to generate other differences between the two regions. The most crucial of these other differences involved the alleged non-occurrence in the heavenly region of any of the physical changes that are ubiquitous and continual on Earth. To reach such a conclusion about the unchangeability of the heavenly bodies, the Aristotelians combined the doctrine of two natural motions with another doctrine, about change and contrariety: that change does not occur unless there is contrariety. They argued that there is no contrariety in the heavens, because the natural motion of heavenly bodies is circular, and natural circular motion has no contrary; but there is contrariety among terrestrial bodies, because their natural motion is two-fold, straight downwards and straight upwards; therefore, heavenly bodies are unchangeable and terrestrial bodies are changeable.

Galileo criticizes the doctrine of two natural motions as conceptually incoherent. One objection is that from the point of view of geometrical simplicity, there is a third simple line, that of the cylindrical helix or spiral, and yet Aristotle ignores it. Moreover, even if we limit ourselves to straight and circular lines, any straight line is simple regardless of whether it intersects the center of the universe; and similarly, circular motion around any point is simple, even if that point is not the center of the universe. Third, it is arbitrary to equate the spontaneous downward motion of heavy bodies with motion toward the center of the universe, rather than toward the center of the Earth; equally arbitrary is to equate the spontaneous upward motion of light bodies with motion away from the center of the universe; and the same applies to the apparent motion of heavenly bodies. Finally, there is an ambiguity in the Aristotelian concept of natural motion, insofar as it has two incompatible meanings: potentially everlasting motion, as in the case of heavenly bodies; and spontaneously free motion, as in the case of terrestrial bodies.

In any case, Galileo objects, the contrariety argument is self-contradictory. If you accept it, then you could give an analogous argument showing the opposite—that the heavenly bodies are changeable. Such an analogous counter-argument could be stated as follows: heavenly bodies have contraries, since they are unchangeable, but terrestrial bodies are changeable, and changeability and unchangeability are themselves contraries; now, all bodies that have contraries are changeable, since change does not occur unless there is contrariety; it follows that heavenly bodies are changeable.

If these criticisms seem abstract, keep in mind that here Galileo is examining the theoretical arguments for the Earth–heaven dichotomy, and so he is trying to expose their theoretical and conceptual flaws. Moreover, this theoretical criticism is made in addition to the observational criticism of the Aristotelian observational argument.

Second Day: Diurnal Motion

Most of the Second Day consists of a lengthy critical examination of the many arguments against the Earth's daily axial rotation. However, it begins with a very brief discussion elaborating the simplicity argument in favor. This argument was common and well known, not only since Copernicus but also in earlier times. We considered it earlier (Chapter 3). Recall that there were two main ways in which terrestrial rotation was widely recognized to be simpler than universal geocentric revolution: the former involves (thousands) fewer moving parts, and only one direction of motion (eastward) rather than two (eastward for planetary revolutions, but westward for the diurnal motion). The main novelty of Galileo's discussion is the clarity and explicit awareness that this argument is not conclusive but merely probable, and the addition of a new and significant simplification embodied in the Copernican system compared to the geostatic one. Let me explain.

Galileo's general formulation of the simplicity argument is as follows. He states that it is more probable that the Earth rotates, rather than the heavenly bodies rotating around it, because while apparent diurnal motion can be explained by either the Copernican or the Ptolemaic hypothesis, given the principle of relativity of motion, the Copernican one is simpler for at least eight reasons, and nature usually operates by the simplest means.

Let's focus on the third of Galileo's reasons for the greater simplicity of the Copernican explanation of diurnal motion: that the periods of revolution have a uniformity in the Copernican system which they lack in the geostatic one. To show this, Galileo begins with a claim which I call the *law of revolution*: it is probably a general law of nature that, whenever several bodies are revolving around a common center, the periods of revolution become longer as the orbits become larger. He then supports this by the well-known fact that the planets revolve in accordance with this pattern, and by his own discovery that Jupiter's satellites also follow the pattern. The important point here is that, although this feature of planetary revolutions was known to the Ptolemaics and incorporated into their system, before the discovery of Jupiter's satellites it would have been rash to generalize a single case into a general law; however, the completely different and unexpected case of Jupiter's satellites suggested that this was not an accidental coincidence but had general systemic significance.

Given the law of revolution, Galileo goes on to point out that, whereas the Earth's diurnal motion in the Copernican system is consistent with the law of revolution, the diurnal motion of the universe in the Ptolemaic system is not. This is because in the geostatic system the diurnal motion corresponds to the revolution of the outermost sphere (whether stellar sphere or *primum mobile*) around the central Earth, but this outermost sphere involves both the largest orbit and the shortest period. So the Copernican system has a uniformity or regularity lacking in the Ptolemaic system.

Let us now go on to the arguments *against* terrestrial rotation. These were arguments that had accumulated for two millennia, starting with ancient thinkers such as Aristotle and Ptolemy, and being repeated and updated by such great contemporaries of Galileo's as Tycho, and by lesser contemporaries such as Johann Georg Locher and Scipione Chiaramonti. In the Second Day, Galileo sometimes quotes explicitly from their works (especially from Aristotle, Locher, and Chiaramonti), sometimes indirectly attributes an argument to them (especially with regard to Ptolemy), and sometimes is silent about the source (especially for the case of Tycho). However, Galileo is always concerned to elaborate, clarify, and strengthen their arguments before refuting them.

Some of their arguments were philosophical, specifically epistemological, the best example being the objection from the deception of the senses. Recall that this argument claimed that if the Earth rotated, our senses would not be telling us the truth. For example, our eyes clearly see falling bodies move vertically, but on a rotating Earth such an observation would mean that their motion was actually slanted eastward; our sense of touch perceives no constant westward wind, which would have to exist on a rotating Earth; and our kinesthetic feeling of rest would be mistaken if the Earth were rotating.

Galileo denies that such cases are genuine instances of deception of the senses.

First, it would be no deception of the senses if on a rotating Earth we perceived bodies falling vertically and failed to perceive any lateral component of their motion due to terrestrial rotation. For motion exists only relative to things that do not share it, and so motion shared by an observer and an observed object does not exist for the observer and is imperceptible to him. But on a rotating Earth, the eastward component of the motion of the falling body would be shared by the observer, and hence it would not be there to be perceived. Note that here Galileo is exploiting the principle of the relativity of motion to elaborate his criticism.

Second, Galileo objects that there would be no wind deception on a moving Earth because there would be no wind for us to perceive; wind is, by definition, air moving relative to the observer, and on a moving Earth the air as well as the observer would be carried along.

Third, our inability to feel the Earth's motion is not a deception either. For our experience with navigation shows that we can feel only changes of motion and not uniform motion, and so the Earth's constant rotation is not something susceptible of being felt. It would be improper to speak of deception in this case.

Besides such a refutation of the premises of the deception argument, Galileo also elaborates a more general criticism that undermines the validity of the argument; such a flaw would be present even if the cases just considered were indeed deceptions, or other cases were found. He points out that if and to the extent that there would be sensory deception on a moving Earth, that would be no reason to conclude that knowledge is impossible; the more correct conclusion would be that knowledge is difficult, that it cannot rely solely on the senses, and that reason plays an equally crucial role.

Here Galileo elaborates a methodological principle that may be called critical empiricism and may be contrasted with the naïve empiricism of the Aristotelians in general. They make the acquisition of knowledge so dependent on sensory experience that if the senses are not completely reliable, then there is no reliable guide in the search for truth, and knowledge is impossible. By contrast, Galileo stresses that if the senses are not always reliable, then we should learn to distinguish situations in which they are reliable from situations in which they are not, and this task can only be performed by reason. So if and to the extent that there is a deception, it would be a deception of reason, the reason of those who from the fact that certain things appear in a certain way conclude mechanically and uncritically (and incorrectly) that they are really that way.

The heart of the Second Day is Galileo's critique of the many mechanical objections, which were based on the observed behavior of falling bodies, of projectiles, and of bodies undergoing whirling and extrusion, and on the principles of Aristotelian physics concerning natural motion, violent motion, and the motion of simple bodies. Galileo's criticism usually involves a combination of an exposure of flaws in reasoning, a correction of observational claims, and an articulation of a new and better physics based on such principles as conservation, superposition, and relativity of motion. One of the most instructive examples of these critiques deals with the anti-Copernican argument based on the ship's mast experiment, and we will look at it shortly. Two other important examples, to be discussed later, involve the objection from vertical fall, which is a related but distinct argument, and the objection from the extruding power of whirling.

As we saw earlier, the ship's mast experiment amounts to dropping a body, such as a rock or ball, from the top of a ship's mast. The experiment is to be done both when the ship is motionless and when it is advancing forward. The experimental claim was that on a moving ship the rock falls to the deck away from the foot of the mast, toward the back of the ship. And the argument was that on a rotating Earth, a rock dropped from the top of a tower would land away from the foot of the tower, toward the west; and since this does not happen, the Earth does not rotate.

Galileo's main criticism of this argument is that its key premise is false: on a moving ship, the rock is not left behind but rather falls at the foot of the mast, the same as it does on a motionless ship. He has two reasons for this. One is an experimental report; the other is an indirect theoretical argument. In this passage of the *Dialogue*, he elaborates only the theoretical argument, whereas the experimental report is found only in his "Reply to Ingoli" (1624).[4] However, the argument in this passage is theoretical not in the sense of being *a priori*, but rather in the sense that its empirical conclusion is based partly on more easily observable phenomena, partly on more easily ascertainable facts, and partly on some theoretical claims that are not arbitrary but can be justified in various ways.

This theoretical argument may be reconstructed as follows. The more easily observable phenomena are that: (1) the undisturbed downward motion of bodies on an inclined plane is accelerated; and (2) their undisturbed motion up an inclined plane is decelerated. The more easily ascertainable fact is that (3) the cause of projectile motion is not the motion of the surrounding air. The reasons for this are that: (3a) wind moves cotton more easily than rocks, and yet rocks can be thrown farther and more easily than cotton; (3b) lead pendulums oscillate much longer than cotton ones; (3c) if the force cannot be impressed directly by the thrower to the projectile, then it cannot be impressed directly by the thrower to the air; and (3d) arrows can be shot against the wind.

From (1) and (2) one may infer that (4) the motion of bodies on a horizontal plane is conserved if undisturbed, and consequently that (5) the horizontal motion which the rock has before being dropped on the moving ship continues even after being dropped, if undisturbed. Now from (3) one can infer that (6) the cause of the motion of projectiles is the impulse conveyed to them by the projector, and consequently that (7) the cause of the horizontal motion of the rock, after it has been dropped, is the horizontal impulse given to it by the hand holding it before dropping. But (8) there is no way in which this horizontal impulse could be disturbed by the vertically downward tendency due to the weight, because (9) the two are not opposed, but are at right angles, to each other, and (10) they have distinct causes (the projector and gravity, respectively). It follows that (11) the horizontal motion of the dropped falling rock is undisturbed, and hence that (12) that motion will continue, and therefore that (13) the rock will end up at the foot of the mast on the moving ship.

This argument is ingenious and plausible, although not completely conclusive and compelling. It presupposes the principle of the superposition of motions, which is lurking around in the justification of proposition (8) by (9) and (10), and which needs more elaboration; and it presupposes the principle of conservation of motion, a version of which is stated in proposition (4). So it is not surprising that Galileo would seek a direct experimental test of the claim that on a moving ship the dropped rock still falls to the foot of the mast, thus refuting empirically the key premise of the anti-Copernican argument based on the ship's mast experiment. Nor is it surprising that Galileo goes on to elaborate and justify these two principles in the next section of the Second Day. Finally, needless to say, Galileo is well aware that his critical argument merely refutes the anti-Copernican objection, but does not provide support for the Earth's motion.

Third Day: Annual Motion

The Third Day considers the annual motion and also has a two-fold focus: a part that elaborates some arguments in favor of attributing this motion to the Earth, and a longer part criticizing arguments against it (that is, in favor of attributing such a motion to the Sun instead).

One of the arguments in favor of the Earth's heliocentric revolution is, as we have seen, based on retrograde planetary motion, claiming that the geokinetic explanation of this phenomenon is more coherent and better than the geostatic explanation. This was a well-known argument since Copernicus, and Galileo adapts it from him. Another of the arguments was partly old and partly new. It was based on the heliocentrism of planetary motions, that is, the claim that the planets (Mercury, Venus, Mars, Jupiter, and Saturn) revolve in orbits centered at the Sun, rather than at the Earth. This had become a widely accepted claim, especially since Tycho, whose system combined it with the traditional geostatic thesis: the Earth stood still at the center of the universe, but the Sun with the whole Solar System of the planets revolved daily and annually around the motionless central Earth. Although Galileo presented some new evidence in support of planetary heliocentrism, he also devised a new argument on that basis in favor of the Earth's annual motion.

The heliocentrism of the five planets says nothing about the motion or rest of the Earth. However, it does say something about the location or position of the Earth: the Earth is located between Venus and Mars, because Venus (and Mercury) are never observed to be on the opposite side of the Sun in the sky, whereas Mars (and Jupiter and Saturn) are sometimes in opposition; thus, the orbit of Mars encloses the Earth (as off center) as well as the Sun (at the center), but the orbit of Venus encloses only the Sun but not the Earth. And then Galileo advances three reasons for the Earth's motion: it is more fitting to have the center (the Sun) rather than a point off center (the Earth) be motionless; the Earth is positioned between two other bodies (Venus and Mars) which perform orbital revolutions; and the period of the Earth's orbital revolution (one year) would be intermediate between the periods of Venus and Mars (nine months and two years, respectively), just as the size of its orbit would be intermediate between the sizes of theirs. This is a kind of simplicity argument, ingenious and plausible; but of course it is not conclusive.

Finally, there is an argument in the Third Day that is novel with regard to both the evidence used and the reasoning arising from it. The new evidence analyzed by Galileo concerns the annual pattern of sunspot paths. And the reasoning is basically that this phenomenon is better explained geokinetically than geostatically.

The observational evidence about sunspot paths had been described in detail in a 1630 book published by Jesuit astronomer Christoph Scheiner, but he gave a geostatic explanation of the phenomenon. This was the same Scheiner with whom in 1612–13 Galileo had had a bitter dispute over the priority of discovery and the interpretation of sunspots in general, although at that time neither had yet observed the annual cycle of sunspots. Although Galileo says in the Third Day that he had observed this annual cycle before and independently of Scheiner's book, Galileo had never mentioned the phenomenon before; moreover, there is no question that he was led to understand its Copernican significance by Scheiner's book. This is the important point for us here.

Galileo summarizes the observable facts about the trajectories followed by sunspots by means of some diagrams, which may be adapted with slight modifications as follows. Figure 15[5] represents four views of the solar disk and of sunspot paths at roughly three-month intervals. MN is a line perpendicular to the line of sight of a terrestrial observer and located in the plane of the ecliptic, which plane we can imagine to be perpendicular to the plane of the paper, while we regard the observer to be located in this plane. Most of the time, the trajectories are both inclined and curved with respect to the plane of the ecliptic.

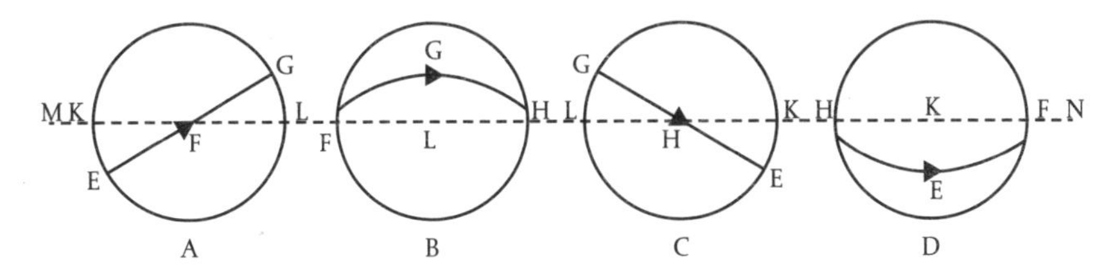

Figure 15. Annual cycle of sunspot paths

However, twice a year, at about six-month intervals, the curvature of the sunspot trajectories is absent and their inclination is maximum, so that sunspots appear to follow a slanted rectilinear path (views A and C). In the first of these views (A), the slant is upwards; a spot that first becomes visible at E appears to move along line EG toward G. In the third view (C), the slant is downwards, with spots first visible at G appearing to move toward E along line GE.

Moreover, the curvature is maximum at two other times of the year (views B and D), also separated by six-month intervals from each other, but interspersed at three-month intervals with the two other views (A and C). At one of these times of maximum curvature (B), the curvature is upward from the plane of the ecliptic, but six months later (D), the curvature is downward. And at these times there is no slant—the spots are seen to begin and end their path at points equally distant from the ecliptic (near F and H in view B; and near H and F in view D). At these times, the direction of motion is the same as for all other times, namely from left to right of the observer.

Aside from these four special times, the paths of sunspots usually appear both curved and slanted and are continuously changing, with the curvature being upward for half a year and downward for the other half, and with the slant also alternating in a similar manner.

Galileo's geokinetic explanation of such sunspot paths is best stated in terms of another diagram (Figure 16).[6] This represents the Earth's annual orbit ABCD, or ecliptic, around the central Sun, represented by the sphere KELG. (The Earth's daily axial rotation is not represented here, and plays no role in this part of the argument. And although the Earth's orbit is drawn in

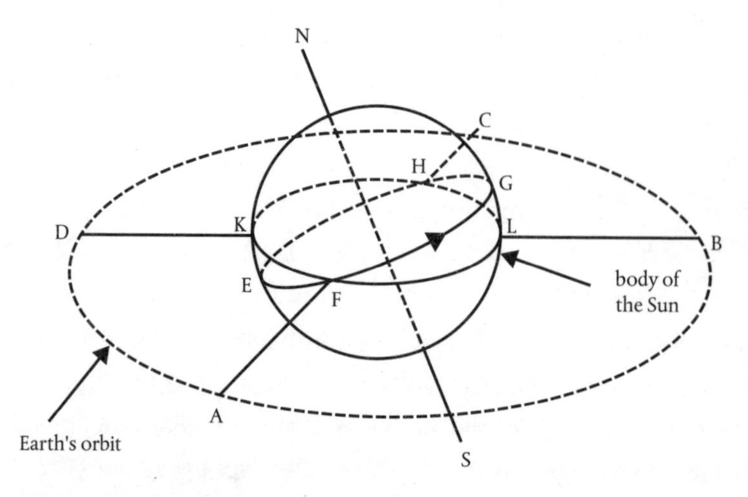

Figure 16. Geokinetic explanation of annual cycle of sunspot paths

an elliptical shape, this is not a reference to Kepler's first law of planetary motion, but reflects the convention that we are looking at a circle from an angle.) AC and DB are diameters of the Earth's orbit. NS is the axis of the Sun's monthly rotation, N corresponding to its north pole, and S corresponding to its south pole; note that axis NS is inclined to the plane of the ecliptic. Circle EFGH is the solar equator, equidistant from the poles, and perpendicular to the axis. KFLH is a great circle of the solar sphere, at the intersection of the solar body and the ecliptic plane.

As the Earth revolves around the Sun in a counterclockwise direction (through points A, B, C, D), terrestrial observers get different views of the path of sunspots. When (from D) the Sun's north pole is tilted toward us, sunspots rotating along the solar equator (or circles parallel to it) will appear to bend southward (i.e., downward), along curve HEF; whereas from B, when it's the Sun's south pole that is tilted toward us, sunspots (still rotating counterclockwise over the solar body) will appear to bend northward (i.e., upward), along curve FGH. On the other hand, about a quarter of a circle after D, from A, we will see the greatest inclination northward (along EFG) but no bending or curving; and from a point reached six months after A, namely from C, the path will again appear straight, still left to right or counterclockwise, but now inclining southward. These visualizations become easier if we compare Figure 16 with Figure 15 embodying the observed path of the sunspots across the Sun (note that the letters in the two figures correspond).

In short, the Earth's annual motion around the Sun explains in a simple, coherent, and natural manner why we observe sunspot paths to exhibit an annual cycle. By contrast, to explain the phenomenon geostatically, the Sun would have to be attributed an additional motion, besides its diurnal and annual motions and monthly axial rotation. Such a fourth motion would have to be one of the following: if the Sun is rigidly attached to its own orb in accordance with the traditional geostatic system, then there must be an annual precession of the solar axis of monthly rotation in a direction (clockwise) opposite to its rotation, to ensure that the north pole of this axis points in the proper sequence relative to a terrestrial observer—left, away, right, and toward; or if the traditional orbs are discarded and solar rotation is conceived in a Galilean manner as an inertial phenomenon, then the Sun's axis must have a daily precession in the same direction as the diurnal motion, to ensure that in the course of a single day the solar axis points always in the same direction relative to the terrestrial observer, and that the pattern of the annual cycle is not repeated on a daily basis. Such a fourth solar motion would increase complexity; it would be *ad hoc*; and it would be inexplicable.

As I mentioned, the Third Day devotes more space to the criticism of the anti-Copernican arguments than to the presentation of the pro-Copernican ones. But here we will have to reverse that space allocation and deal with the objections more briefly.

One group of objections to the Earth's annual motion had already been answered 20 years earlier, at the time of the original telescopic discoveries. This is the case, for example, for the arguments based on the non-observation of Venus's phases and of the eight-fold variation in Mars's apparent diameter. As we saw earlier, these arguments are unanswerable as long as one is limited to naked-eye observations, but are easily answered once the telescope reveals the phenomena in question. Galileo had never found the occasion to discuss these issues explicitly and systematically, and he does so in the Third Day.

Galileo makes a methodological confession in this passage to the effect that he personally judged these observational arguments to be so strong that, without the telescope, he (unlike Copernicus) could have never taken the Earth's motion seriously. In memorable words uttered by Salviati, Galileo reveals:

> These are so clearly based on our sense experience that, if a higher and better sense than the common and natural ones had not joined with reason, I suspect that I too would have been much more recalcitrant against the Copernican system than I have been since a lamp clearer than usual has shed light on my path...These are the difficulties that make me marvel at Aristarchus and Copernicus; they...were unable to solve them; and yet...they trusted what reason told them so much that they confidently asserted that the structure of the universe can have no other configuration but the one constructed by them.[7]

Finally, Galileo also discusses in the Third Day an objection which he could not really refute: the argument from the apparent position of fixed stars, also known as the argument from annual stellar parallax. Recall that this objection argued that if the Earth revolves around the Sun, then in the course of a year the apparent position of any one fixed star (measured, for example, by its angular elevation above the ecliptic) should change; in other words, there should be an annual stellar parallax. But no such parallax was observed, even with the telescope. Galileo's lengthy discussion amounts to a clarification of the argument, an analysis of the relevant issues, and a description of a research project designed to detect the parallax if it exists.

In fact, the anti-Copernicans had many confusions and misconceptions about exactly what the stellar changes should be. Galileo argues constructively that the Earth's heliocentric revolution should cause changes such as the following (Figure 8, in Chapter 3): stars lying on the ecliptic plane would exhibit a change in apparent magnitude but no change in angular elevation; stars near the pole of the ecliptic would show a change in elevation but no change in magnitude; stars located in-between (the pole and the ecliptic) would display both types of changes; the amount of such changes would depend on the angular position of a star, such that lower positions would cause bigger changes in magnitude and smaller changes in elevation; finally, the amount of both such changes would vary inversely with the distance of a star. Moreover, because of the extremely large distances involved, all such changes would be extremely small.

Galileo also points out that the objection assumes that the failure to observe annual changes in stellar appearances implies that there are no such changes. And he thinks that this assumption is precisely the weak point of this argument. Perhaps they are not observed because they do not exist, but their absence may also be because no one has searched for them seriously or systematically enough, or with the appropriate instruments, or with the necessary skill.

Accordingly, this passage also contains some practical suggestions about observational procedures and techniques, designed to detect such changes in stellar appearances. Galileo says that we need instruments whose sides are miles long, so that differences of seconds in stellar elevation correspond to distances of the order of cubits along the instrument. This would contrast with the capabilities of even the very best previous astronomical instruments, such as those of Tycho: in his apparatus, the celestial quantities being measured corresponded to differences of a hairsbreadth on the instrument. Galileo is talking about using topographic features on the Earth's surface, such as mountains. The idea is, for example, to observe the changes in the way a particular star would be hidden by some beam on top of a building, at the top of a mountain, by carrying out observations from the valley below, and at different times of the year.

In this discussion, unlike all other critiques of anti-Copernican arguments, Galileo is not really refuting the argument. Rather, he is admitting that the changes in stellar appearances due to the Earth's heliocentric revolution are not observed, and he is elaborating a research program designed to detect them if they exist. His elaboration consists of a theoretical analysis of the details of such observational consequences, and of practical suggestions for making the observations. Such a research program was indeed followed in part by subsequent astronomers, such as the Englishman James

Bradley, who discovered stellar aberration in 1729; and the German Friedrich Bessel, who measured an annual stellar parallax in 1838.

Fourth Day: Tides

The Fourth Day examines a topic which Galileo had been hinting at in several previous passages—the problem of explaining why tides occur. He makes it clear at the outset that he is going to explain this phenomenon on the basis of the Earth's motion, and that this explanation will provide a novel argument in favor of the Copernican hypothesis that the Earth moves.

Galileo also makes it clear that this tidal argument for the Earth's motion is different from his previous pro-Copernican arguments because it involves physical considerations about terrestrial phenomena, whereas those others have all involved astronomical considerations about the heavenly bodies.

Terrestrial physical phenomena had been extensively discussed in the Second Day, in the context of the criticism of the mechanical objections to the Earth's rotation. But that criticism showed only that the physical phenomena being appealed to could take place on a moving Earth, not that they provide evidence in favor; in other words, that criticism merely refuted the anti-Copernican arguments, without yielding pro-Copernican ones. On the other hand, the tidal motion of the sea is different from the other terrestrial mechanical phenomena: according to Galileo, whereas the latter could occur on both a motionless and moving Earth, the former could *only* occur on a moving Earth; and this is what yields a pro-Copernican argument.

The discussion begins with a description of the most basic facts about the tides. The phenomenon consists of three kinds of basic motions of seawater: up and down motion, such as the vertical rising and falling of the water level visible inside a harbor, along the sides of a pier; back and forth motion, such as the horizontal water currents observable in certain straits or narrows; and a combination of these two, such as the ebb and flow that takes place in some beaches and shallow coastlines, in which the water alternates between both rising and flowing inland, and dropping and flowing outward to sea.

Moreover, these basic tidal motions occur in accordance with various cycles, the most important and noticeable one being the diurnal period: the water usually rises or flows in one direction for about six hours, and then it drops or flows in the opposite direction for another six hours. So during the twenty-four hours of a single day, there are two high and two low tides separated by six-hour intervals, and there are two full cycles of 12 hours during each of which the water exhibits all the motions that characterize a particular location.

This phenomenon had puzzled mariners and thinkers from time immemorial, and various attempts had been made to explain why seawater exhibits such motions. Galileo gives a brief critical review of the main alternative explanations. Some had tried to explain the tides as a result of an attraction exerted by the Moon toward the water directly below. This explanation was in part supported by the well-known and undeniable fact that there is a correlation between the daily motion of the Moon and the tides. Yet Galileo dismissed such an explanation for two reasons. First, lunar attraction would produce the same tidal motion in all parts of a given sea, for example, both in Venice at the northwestern end of the Adriatic Sea and in Dubrovnik at the southeastern end of the same sea; but that is not the case. Second, he regarded the lunar-attraction account as methodologically inappropriate, insofar as attraction was an "occult" property involving a magical view of nature and the resulting explanation would be tantamount to explaining a phenomenon by giving a name to it. At one point, he even chides Kepler for his "childish" belief in a lunar-attraction theory. The irony is of course that the lunar-attraction explanation turned out to be essentially correct; Isaac Newton explained the tides in 1687 in terms of the law of universal gravitation and the different gravitational forces exerted primarily by the Moon (but also by the Sun) on different parts of the oceans and land.

The first step in Galileo's own explanation is to show that when a container of water undergoes acceleration or retardation, the water experiences vertical up and down motions at the extremities and horizontal back and forth motion in the middle part of the container. Labeling such movements tidal-like motions, and taking the term acceleration in the strict general sense that includes retardation as negative acceleration, the first step is to state and support the *generalization that acceleration causes tidal-like motion*.

To see this, consider a boat with a large rectangular tank full of water, going from one seaport to another over a calm sea, such as one of the boats that deliver fresh drinking water to the city of Venice from the nearby coast. Consider now what happens to the water in the tank as the boat begins its journey and starts moving. The acceleration imparted to the boat and the tank is not instantly communicated to the water in the tank. This water will tend to be left behind and flow backwards, so that its level will rise at the back end of the tank and drop at the front end. Being a fluid, the water will then tend to flow in the opposite direction, from the back to the front end of the tank; the water level will then rise in front and drop in the back. This process of oscillation will continue for a while, even though the boat may have reached a uniform cruising speed. Moreover, during this process, if we look at the middle part of the tank, the water there will mostly not rise or drop, but rather move horizontally backwards first, then forward, then backwards again, and so on. After a while, however, if the boat continues moving uniformly over a calm sea, the water in the tank will also calm down and no longer experience those motions within the tank; it will simply follow the forward motion of the boat.

Now suppose the boat runs aground in shallow water, thus experiencing a strong reduction in speed until it stops. Such retardation will cause the water in the tank to start its oscillatory motion, like during the initial acceleration, except in reverse order. At first, the water will rush forward, increasing the level in front and lowering it at the back, then the other way, and the oscillations will continue for a while before dying down. Besides such acceleration on starting and stopping, merely increasing or decreasing the speed of the boat while cruising would have similar effects.

So Galileo's first intermediate conclusion is that the acceleration or retardation of a body of water causes it to undergo tidal-like motions. Now, he applies this generalization to the case of a moving Earth in the Copernican system, making the key analogy that a sea basin, such as the Mediterranean Sea, is a container of water being carried by the moving Earth. And the generalization applies in the sense that the Earth's axial rotation and its orbital revolution combine in such a way as to generate daily accelerations and retardations in the motion of such a sea basin. The second step in the Galilean explanation is precisely to show that, if the Earth were moving (with the two motions of axial rotation and orbital revolution), then every point on the Earth would regularly and alternately undergo a daily acceleration and a daily retardation.

In Figure 17,[8] circle GEC is the orbit in which the Earth moves in a counterclockwise direction, around the central Sun at A. Circle DEFG represents the Earth, which rotates around its own center B, in the same (counterclockwise) direction. Here both motions are taken to be constant and uniform: the Earth's center B moves around the Sun A at a speed that enables it to traverse the whole circumference of circle GEC in one year; while any given point on the Earth, for example D, in the period of 24 hours traverses the whole circumference of the Earth, DEFG.

Now, consider how these two motions combine with each other. For a terrestrial point located on the opposite side from the Sun and experiencing midnight, for example D, the diurnal speed and the annual speed are in the same direction, toward the left; so they add up to give point D an actual speed that is the sum of the two, by reference to absolute space or at least to the center of the Sun A. But for a terrestrial point located on the side facing the Sun and experiencing noon, for example F, the annual and diurnal

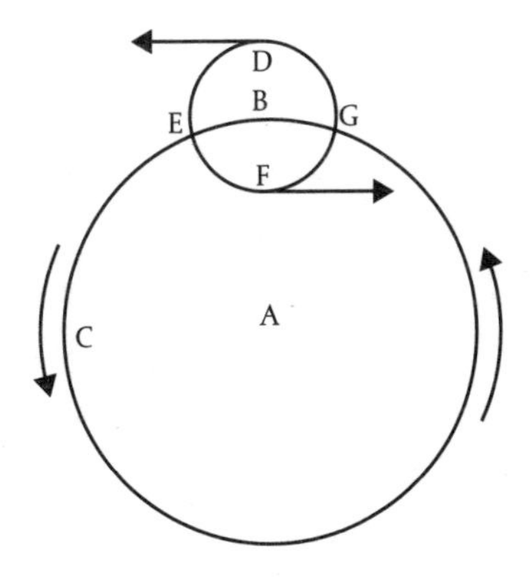

Figure 17. Geokinetic explanation of daily tides

speeds are in opposite directions, the former still toward the left but the latter toward the right; thus they work against each other to give point F an actual speed that is the difference between the (greater) annual speed and the (smaller) diurnal speed. For intermediate midway points on the Earth's surface, such as E and G, their diurnal speed has no effect on their annual speed, at least from the point of view of the west-to-east (i.e., right-to-left) direction; thus, such points move toward the left with a speed equal to just the annual speed.

At D the total actual speed is the maximum; at F it is the minimum. This applies to every point on the Earth's surface in the course of 24 hours, and so any such point will alternate between such a maximum and a minimum speed. Moreover, in moving from D (through E) to F, a terrestrial point is undergoing retardation; whereas in moving from F (through G) to D, it is undergoing acceleration. This also applies to every point on the Earth, and so we may also say that every day every such point experiences acceleration for 12 hours and then retardation for another 12 hours, and so on. Furthermore, while moving from G to D and to E, every terrestrial point moves at a speed that is greater than the annual speed; whereas, when moving from E to F and to G, it moves at a speed that is smaller than the annual speed. So we may also say that every day every point on the Earth alternates between moving at a speed greater than the annual speed for 12 hours, and moving at a lower speed for the next 12 hours, and so on.

The daily accelerations and retardations just described constitute for Galileo the *primary cause* of the tidal motions of the sea. That is, he means partly that without such changes of speed there would be no tides; in this sense, they provide a necessary condition for the tides. He also means that such changes of speed get the whole process of the tides started; and in this sense, they contribute to the production of the tides. However, by itself the primary cause is not sufficient, and we need other concomitant causes, which he subdivides into secondary and tertiary causes. The secondary causes involve the fluid properties of water, which Galileo elaborates at length. The tertiary causes are such factors as the flow of large rivers into small seas, the action of winds, and the topographical interrelationships of sea and land.

Other complications stem from the fact that, besides the basic tidal motions, the phenomenon exhibits various features, secondary tidal effects, that also require explanation; an example being that there are no tides in lakes and small seas. Finally, besides the daily period of the tides, Galileo also tries to explain the monthly and annual variations in the tides.

It is obvious that the basic structure of Galileo's tidal argument is explanatory, in the sense that its conclusion (the Earth's motion) is supported by showing that it can provide an explanation of some facts (the tides) stated in the premises. In this regard, this argument is structurally similar to several others presented earlier in the book that provide explanations of various astronomical facts (some old and some new). However, in those previous arguments a key point was that the geokinetic explanation was simpler or more coherent than the geostatic. Instead, in the present case, Galileo believes and argues that there is *no* geostatic explanation; that there is no way of explaining the tides as long as the Earth stands still. So the complexities of his own geokinetic explanation do not really weaken it; they are needed because of the complexity of the phenomenon to be explained.

On the other hand, Galileo is aware that just because no satisfactory geostatic explanation of the tides has yet been devised does not mean that no such explanation is logically possible. Such a possibility cannot be ruled out; this is so partly for theological reasons stemming from divine omnipotence, but also for methodological reasons stemming from the logic of explanation. And Galileo mentions the theological argument at the very end of the Fourth Day, as required by the pope and the censors; and he elaborates the logic and methodology of explanation in several previous passages. Indeed, we can interpret the later elaboration of a gravitational explanation of the tides by Newton as providing an alternative and better explanation than Galileo's, and thus weakening and invalidating his tidal argument in a straightforward manner.

Rhetorical Flourishes and Poetic License

The account just given provides a summary of the essential parts of the content and structure of the *Dialogue* from the point of view of astronomy, physics, and the Copernican controversy. I have also summarized a few of Galileo's methodological or epistemological discussions that happened to be especially relevant to the argument and particularly incisive or memorable. Indeed, the book is full of such concrete philosophical reflections, so much so that they make it easy to see why, as I put it in the Introduction, Galileo may be rightly regarded as the Socrates of methodology and epistemology.

There is another aspect of the book which I have so far almost completely ignored. I refer to the book's rhetorical dimension. Here, by rhetoric I mean the theory and practice of verbal communication, in all its variety of expression.

The book's rhetoric derives in part from its universalist aim; Galileo is addressing several audiences at once. It derives in part from its dialogue form, which means that there is a certain amount of drama unfolding before the reader. It stems from the controversial character of the scientific and epistemological issues discussed, which means that we are witnessing a brilliant polemic. It also arises from the context of Galileo's struggle with the Catholic Church: in writing the book he was taking considerable risks and could not always safely say what he meant or mean what he said. Finally, the rhetoric originates to some extent from the fact that the practice of science at that time was socially and financially dependent for the most part on the patronage of princes; this means generally that Galileo's career was partly that of a courtier, and specifically that his book represented an action in an intricate and delicate network of patronage involving the Tuscan Medici court in Florence and the Vatican court of Pope Urban VIII in Rome.

I am *not* equating rhetoric with the art of deception, in particular the skill of making the weaker argument appear stronger. So understood, rhetoric would be an inherently objectionable activity, whereas my definition allows both good and bad rhetoric. Nevertheless, rhetoric does not easily mix or coexist with scientific inquiry, critical reasoning, and methodological reflection; it considerably complicates the proper understanding and evaluation of the book. Indeed, the book's rhetoric is one of the things that led to the Inquisition proceedings against it, as we shall see.

The *Dialogue* also happens to be a great work of art, because Galileo was a gifted writer who poured his heart and soul into this work, so much so that many passages achieve a high degree of literary and aesthetic value. Indeed, Galileo is widely regarded to be one of the greatest writers in the (800-year) history of Italian literature, rivaled only by poet Dante Alighieri (1265–1321) and his *Divine Comedy*. If we define poetry in terms of expressiveness and imagery, rather than formalistically (in terms of verse, rhyme, and meter), we could say that the *Dialogue* is a scientific and philosophical poem. The book is full of passages which are best understood and appreciated in terms of artistic inspiration, aesthetic imagination, and poetic license. This aspect of the book also played a part, however small, in the trial proceedings, specifically in Galileo's confession of some wrongdoing, stemming from literary vanity. Galileo is in places prone to rhetorical excess and poetic license.

We have already seen that the three speakers in the book's dialogue are named Simplicio, Salviati, and Sagredo, and that they represent, respectively, the Aristotelian, Copernican, and layperson's point of view. In the book's preface, Galileo states that Salviati and Sagredo were names of recently deceased friends with whom he had discussed such topics; but Simplicio was the (Italian) name of a famous commentator to Aristotle's works, who lived in the sixth century AD, and whose Latin name is Simplicius. This seems a reasonable choice. But the name Simplicio also has the connotation of *simpleton*. Moreover, recall that Simplicio is the interlocutor who loses all the arguments (although he is neither stupid nor simple-minded, and occasionally displays flashes of great logical insight,[9] which are shared by Salviati and Galileo himself). Hence, with the name Simplicio, Galileo seems to be engaged in a *double entendre*, to remind readers of the inferiority of the Ptolemaic system to the Copernican.

Take an example involving the Earth–heaven dichotomy. This thesis included the idea that the heavenly bodies (primarily the Sun, but also the planets and the fixed stars) cause physical changes in the terrestrial bodies, although they are themselves unchangeable. We saw earlier that Galileo criticizes the Earth–heaven dichotomy on substantive observational grounds, as well as on theoretical conceptual grounds. But he also could not resist the temptation of making fun of this particular idea. In the middle of the First Day, he points out that this is like "placing a marble statue beside a woman and expecting children from such a union."[10]

Other aspects of the Ptolemaic system also lent themselves to ridicule. At the end of the Second Day, Galileo picks on its feature that the Earth is at the middle of everything, even though the heavens are noble, perfect, and incorruptible and the Earth is base and full of imperfections and impurities: "What a way of separating the pure from the impure and the sick from the healthy—to give those who are infected room at the heart of the city! And I thought that the pesthouse should be located as far away as possible!"[11]

Other types of rhetoric abound. For example, we have seen that some of the objections to the Earth's diurnal motion were based on mechanical phenomena and the laws of motion. Some of these mechanical arguments dealt with the range and accuracy of gunshots in various directions on a rotating Earth, comparing and contrasting gunshots toward the east and west, toward the north and south, vertically upwards, and in a horizontally point-blank direction. Such gunshot arguments are not found in Aristotle's works, since gunpowder had not been invented during his lifetime. However, after this invention, anti-Copernicans formulated new arguments against the Earth's motion based on gunshots.

Now, at the beginning of the Second Day, Simplicio confesses his relative ignorance about such modern arguments, and Salviati volunteers to give full and detailed statements of them. After Salviati is finished, Simplicio expresses his joy at the fact that the truth (that the Earth stands still) can be supported by such irrefutable arguments and novel evidence (as compared with that available at the time of Aristotle). At this point, Sagredo interjects:

"What a pity that there were no cannons in Aristotle's time! With them he would have indeed conquered ignorance."[12] The rhetorical service performed by this witticism is to insinuate that the anti-Aristotelians have such an understanding of the Aristotelian position as to be able to eloquently present new evidence in its favor, and thus to incline readers to think more highly of Copernicanism.

The Third Day does not lack its share of rhetoric. Here is a different kind of example. In the discussion of the heliocentrism of planetary motions, Galileo hurls an insult to some opponents of Copernicanism by calling them "men whose definition contains only the genus but lacks the difference."[13] In traditional logic, definitions were given by identifying the genus and the species (or specific difference) to which the thing to be defined belongs; the genus is a broader category of classification, and the species (or specific difference) is a subdivision within the genus. Modern biological taxonomy still follows this procedure. Modern humans, *Homo sapiens*, belong to the genus *Homo* and the species *sapiens*.

Now, traditional Aristotelian doctrine defined man as "rational animal," namely as belonging to the genus "animal" and the species (or specific difference) "rational." If from this definition of man the species is removed, we are left with "animal"; so a person whose definition contains only the genus is an alleged rational animal who is not really rational but only a mere animal. Galileo is here engaged in name-calling; subtle and not prosaic to be sure, but name-calling nonetheless.

Finally, here is an example from the Fourth Day. We saw earlier that, before elaborating his own geokinetic explanation of the tides, Galileo gives brief criticisms of alternative explanations. One of these explained the phenomenon as due to the Moon, and one version of the Moon theory claimed that, as the Moon moved above a particular sea on Earth, the Moon's heat warmed the water below, causing it to expand and its level to rise. One of Galileo's criticisms of this version is the empirical one of inviting anyone to test the temperature of water at high and at low tides, and see that there is no difference. To this he adds, referring to proponents of this theory, the following gem: "tell them to start a fire under a boiler full of water and keep their right hand in it until the water rises by a single inch due to the heat, and then to take it out and write about the swelling of the sea."[14] He does not elaborate, but I suppose the point is that the expansion of water due to heat is not that great, so anyone trying the boiler experiment would burn their hand waiting for the water to rise by an inch.

CHAPTER 7

THE INQUISITION TRIAL (1632–1633)

Complaints about the *Dialogue*

Galileo's *Dialogue on the Two Chief World Systems, Ptolemaic and Copernican* was published in Florence in February 1632. It was well received in scientific circles. But a number of rumors and complaints began emerging and circulating among clergymen and officials, especially in Rome.

The most serious complaint was that the book violated a precept issued by the Inquisition to Galileo in 1616; this was the special injunction to stop holding, defending, or teaching the Earth's motion in any way whatever. Recall that the Inquisition proceedings contain a notary memorandum dated February 26, 1616 which reads like a report of what happened on that day to implement the orders of Pope Paul V from the Inquisition meeting on the previous day. The document states that, first, Cardinal Bellarmine gave Galileo the informal warning to abandon Copernicanism (that is, to stop holding or defending the Earth's motion as true or as compatible with Scripture); and immediately afterwards, Commissary Seghizzi issued Galileo the formal injunction. The charge was that Galileo's just published book was a clear violation of this special injunction, for, whatever else the book did, and however else it might be interpreted, it undeniably taught, supported, and defended the Earth's motion, at least as a hypothesis; this transgression was aggravated by the fact that he had not disclosed this injunction to the officials who authorized the book's publication, and so he could also be accused of deception.

To be sure, the memorandum did not bear Galileo's signature, and this alone was suspicious. In fact, as we know from our earlier discussion, the factual accuracy of this document and the legal validity of the precept were questionable, since they conflict with all other documents and events. However, in 1632, some of these other documents were unknown or unavailable, and in any case such anti-Galilean critics had not noticed any conflicting evidence, but focused on this particular incriminating item. Under different circumstances such legal technicalities could have been taken seriously. However, too many other difficulties were being raised about the book.

One of these other difficulties was that the book only paid lip service to the stipulation about a hypothetical discussion, which represented Pope Urban's (and Bellarmine's) compromise. In reality, the book allegedly treated the Earth's motion not as a hypothesis, but in a factual, unconditional, and realistic manner, as a factual possibility or probability about physical reality.

This was a more or less legitimate complaint. But the truth of the matter is that the concept of hypothesis was itself ambiguous at that time and in that context.[1] By hypothetical discussion, Urban meant treating the Earth's motion merely as an instrument of prediction and calculation, rather than as a potentially true description of reality. On the other hand, Galileo took a hypothesis to be an assumption about physical reality, which accounts for what is known, and which may be true, though it has not yet been proved to be true.

Then there was the complaint that the book did not show the proper appreciation of divine omnipotence. Remember that this was the basis of Pope Urban's favorite (and allegedly irrefutable) argument against the Earth's motion, and that Galileo had complied with the request to include this argument in the book. The charge was that the argument is uttered by Simplicio, who is allegedly a fool, while the other two speakers do not express a sufficiently favorable assessment of the argument; and that the argument is not prominently displayed in the book insofar as it occurs in a passage (on the last page) that is hard to find and easy to miss.

As I pointed out earlier, Galileo did genuinely (and correctly) believe that there was something right about the divine-omnipotence argument.

Aspects of some formulations of the argument are correct, insofar as they reflect the logic and methodology of theoretical explanation, which Galileo discusses in other passages. However, other formulations are easily answered, for example, by pointing out that a corresponding objection applies to the geostatic world view as well. The root problem was that Galileo's book was also supposed to steer clear of theological discussions, and so on this topic he did not have viable options. This charge, then, was a plausible but debatable one.

There were also complaints involving alleged irregularities in the various permissions to print which Galileo obtained. For example, besides displaying the imprimatur for the city of Florence, the printed book displayed an imprimatur for the city of Rome. In fact, originally he had planned to have it published in Rome, but then there was an unavoidable change of venue, due partly to the death in 1630 of Galileo's patron, Prince Federico Cesi, founder and head of the Lincean Academy, who was going to sponsor its publication. The change was also due to an outbreak of the plague in 1630–2, which made it almost impossible for the book manuscript to be sent back to Rome, after Galileo had returned home to Florence.

To add to all this, there were substantive criticisms of various specific points discussed in the book. Feelings had been hurt by some of Galileo's rhetorical excesses and biting sarcasm, and malicious slanders circulated, suggesting that the book was in effect a personal caricature of the pope himself. One of these slanders stemmed from Pope Urban's favorite argument, the divine-omnipotence objection to Copernicanism, being put in the mouth of Simplicio. Aside from the alleged failure to appreciate divine omnipotence, as we saw above, Galileo was thought to be insulting Pope Urban by portraying him as a simpleton.

This complaint could be cleared up by pointing out that "Simplicio" was also the Italian name of the sixth-century Aristotelian philosopher Simplicius. Galileo himself had so stated in the book's preface. Moreover, it would be natural and rhetorically proper to have Simplicio advance the divine-omnipotence objection, given that he is the interlocutor who defends Ptolemy and criticizes Copernicus.

Another malicious slander involved the image printed at the bottom of both the book's frontispiece and the title page (see Figure 18). It depicts three fishes in a circular pattern, with the mouth of each fish touching the body of another. One rumor insinuated that this image was mocking the pope's widely criticized "nepotism." In fact, Urban had carried this practice to new heights by appointing several relatives to important positions: his brother Antonio Barberini, Sr., as cardinal, and eventually as secretary of the Inquisition; his nephew Francesco also as cardinal, and eventually as secretary of state; his nephew Antonio, Jr., also as cardinal; and his nephew Taddeo as governor of the city of Rome. Fortunately, this particular slander was easily and conclusively refuted when it was demonstrated that the image of the three fishes in a circular pattern was the long-standing logo of the book's publisher, which usually appeared in all the books it printed.

The sheer number of complaints, and the seriousness of some of the charges, were such that the pope might have been forced to take some action even under normal circumstances. But Urban VIII was himself in political trouble due to his behavior in the Thirty Years' War between Catholics and Protestants (1618–48). Given his especially vulnerable position, not only could Urban not continue to protect Galileo, but he chose to use Galileo as a scapegoat to reassert, exhibit, and test his own authority, power, and religious credentials.

In the summer of 1632, sales of the *Dialogue* were stopped, and unsold copies confiscated. The pope did not immediately send the case to the Inquisition, but took the unusual step of first appointing a special commission to investigate the matter. This three-member panel issued its report in September 1632,[2] and it listed as areas of concern about the book all of the above-mentioned problems, except one—the slander about the image of three fishes, which by then had already been refuted. In fact, it is from the commission's report that we learn about all these complaints that had been accumulating since the book's publication. In view of the report, the pope felt justified in forwarding the case to the Inquisition.

Yet the report can also be read as leaving open the question of whether to prosecute Galileo, or merely have him revise the *Dialogue*. So perhaps Urban

DIALOGO

DI

GALILEO GALILEI LINCEO

MATEMATICO SOPRAORDINARIO

DELLO STVDIO DI PISA.

E Filoſofo, e Matematico primario del

SERENISSIMO

GR. DVCA DI TOSCANA.

Doue ne i congreſſi di quattro giornate ſi diſcorre ſopra i due

MASSIMI SISTEMI DEL MONDO TOLEMAICO, E COPERNICANO;

Proponendo indeterminatamente le ragioni Filoſofiche, e Naturali tanto per l'vna, quanto per l'altra parte.

CON PRI  VILEGI.

IN FIORENZA, Per Gio: Batiſta Landini MDCXXXII.

CON LICENZA DE' SVPERIORI.

Figure 18. Title page of Galileo's *Dialogue*

wanted to convey the impression that the report was unequivocal in recommending prosecution, which is what he said to the Tuscan ambassador to Rome. In other words, it is likely that Urban was mostly manipulating the proceedings, for various reasons. In part, he was probably trying to cover up his own permissivism and complicity in the writing and publication of the *Dialogue*.[3] Urban may also have been acting on his perception that this book came close to formal heresy by its failure to treat the Earth's motion hypothetically and show the proper appreciation for divine omnipotence; however misconceived and incorrect this perception was, it appears to have been genuinely and sincerely felt by Urban.[4] So Galileo was summoned to Rome to stand trial.

Proceedings Begin

The entire autumn of 1632 was taken up by various attempts on the part of Galileo and the Tuscan government to prevent the inevitable. The Tuscan government got involved because of Galileo's position as Philosopher and Chief Mathematician to the grand duke, because the book contained a dedication to the grand duke, and because the grand duke had been instrumental in getting the book finally printed in Florence.

At first, they tried to have the trial moved from Rome to Florence. This request was summarily rejected by the Inquisition, because of the seriousness of the charges and the notoriety of the person involved. Then they asked that Galileo be sent the charges in writing, and that he be allowed to respond in writing. The reply was that the Inquisition did not operate in such a manner. As a last resort, three physicians signed a medical certificate stating that Galileo was too ill to travel. This was basically true: he was 68 years old, physically frail, and afflicted by his share of ills; moreover, the recent outbreak of the plague meant that travelers from Tuscany to the Papal States were subject to quarantine at the border, and this would have been exhausting for such an ill old man. The medical excuse was also rejected. In fact, at the end of December, the Inquisition sent Galileo an ultimatum: if he did not come to Rome of his own accord, they would send officers to arrest him and bring him to Rome in chains.

On January 20, 1633, after making a last will and testament, Galileo began the journey. (I leave it to your imagination to speculate what the connection may have been between these two new-year's resolutions.) Because of an eighteen-day quarantine at the border, the 173-mile trip took 24 days.

On his arrival in Rome, Galileo was not placed under arrest or imprisoned by the Inquisition, but was allowed to lodge at the Tuscan embassy (Palazzo Firenze). However, he was ordered not to socialize and to keep himself in seclusion until he was called for interrogations.

These were slow in coming, as if the Inquisition wanted to use the torment of the uncertainty, suspense, and anxiety as part of the punishment to be administered to the old man. It was very much in line with one reason mentioned earlier by officials as to why Galileo had to make the journey to Rome, despite his old age, ill health, and the epidemic of the plague: it was an advance partial punishment or penance, and if he did this the inquisitors might take it into consideration when the time of the actual proceedings came.

The first interrogation was held on April 12. In accordance with standard Inquisition practice, this hearing was held at the Inquisition palace, in the office of the commissary, who at that time was Vincenzo Maculano;[5] also present were the assessor, the prosecutor, the notary, and one or two assistants. The questions did not focus on Pope Urban's complaint about the book's failure to treat the Earth's motion hypothetically and to appreciate divine omnipotence, but rather on the events of 1616. In answer to various questions, the defendant claimed three main things.

Galileo stated that in February 1616 Cardinal Bellarmine had given him an oral warning, prescribing that the geokinetic theory could be neither held nor defended as true or compatible with Scripture, but only discussed hypothetically. However, Galileo denied having received from the Inquisition Commissary Seghizzi a special injunction not to hold, defend, or teach the Earth's motion in any way whatever; as supporting evidence, he introduced a copy of the certificate which Bellarmine had written for Galileo in May 1616, which said nothing about Seghizzi's injunction. Galileo's third main claim was made in answer to the question why, in light of Bellarmine's warning and/or Seghizzi's injunction, he had not obtained any permission to write the book in the first place, and why he had not mentioned them when obtaining permission to print the book; these omissions had angered the pope and had made him feel deceived. Galileo answered that he had not done so because he only knew about Bellarmine's warning, and the book did not really hold or defend the Earth's motion, but rather showed that the arguments in its favor were not conclusive, and thus it did not violate that warning.

This was a very strong and viable, but not unproblematic, line of defense. Bellarmine's certificate clearly surprised and disoriented the Inquisition officials. But the issue now became whether Galileo's book violated Bellarmine's warning by defending the truth of the geokinetic theory.

From this point of view, it could be argued that in the book, despite its dialogue form, repeated disclaimers that no assertion of Copernicanism was intended, the inconclusive character of the pro-Copernican arguments, and the presentation of the anti-Copernican and pro-geostatic arguments, it was readily apparent that the pro-geostatic arguments were being criticized and the pro-Copernican ones were being assessed favorably. Thus, it could be said that the book was arguing in favor of the truth of Copernicanism, and hence was defending it. This in turn meant that, even from the point of view of Bellarmine's informal warning to Galileo, the *Dialogue* was a transgression.

This complaint was more or less legitimate, but also questionable. For, in this regard, the issue was whether Bellarmine's warning not to defend the truth of Copernicanism included a prohibition to evaluate the arguments, and whether such a prohibition was reasonable. Clearly, Galileo *did* engage in evaluation, besides presentation, interpretation, and analysis; and his evaluative conclusion was that the pro-Copernican arguments were much better than the anti-Copernican ones. But if his evaluation was correct, fair, or reasonable, could he be blamed for the result? In other words, one could say that the *Dialogue* was *discussing*, not defending, the Earth's motion, insofar as it was a critical examination of the arguments on both sides; admittedly, although the pro-Copernican arguments were not conclusive, they were stronger than the anti-Copernican ones, but that was not Galileo's fault.

Out-of-Court Plea-Bargaining

Bellarmine's certificate was obviously new and crucial evidence that could not be ignored. Therefore, it took another two and one-half weeks before the Inquisition decided on the next step in the proceedings. In the meantime, Galileo was not allowed to return to the Tuscan embassy, where he had been lodging since his arrival in Rome in February; rather, he was detained at the Inquisition palace, but allowed to lodge in the chief prosecutor's apartment. What the inquisitors decided was something very close to what might be called an out-of-court settlement involving a plea-bargaining agreement.

They would not press the most serious, but least provable, charge (of having violated the special injunction). Nor would they press the charge of having violated Urban's request for a hypothetical treatment of the Earth's motion and an appreciation of divine omnipotence. However, Galileo would have to plead guilty to the lesser charge of having transgressed Bellarmine's warning not to defend the truth of Copernicanism, in regard to which Galileo's defense, although viable, was the weakest. To reward such a confession, they would show some leniency toward such a lesser violation.

To work out this deal, the Inquisition asked three consultants to determine whether Galileo's *Dialogue* taught, defended, or held the geokinetic theory; in separate reports all three concluded that the book clearly taught and defended the doctrine, and came close to holding it. Then the commissary general of the Inquisition talked privately with Galileo to try to arrange the deal, and after lengthy discussions he succeeded. Galileo requested and obtained a few days to think of a dignified way of pleading guilty to the lesser charge.

Thus, on April 30, the defendant appeared before the Inquisition officials for the second time, and signed a deposition stating the following. Ever since the first hearing, he had reflected about whether, without meaning to, he might have done anything wrong. It dawned on him to reread his book, which he had not done for the past three years, since completing the manuscript. He was surprised by what he found, because the book did give the reader the impression that the author was defending the geokinetic theory. This had not been his intention. Galileo attributed it to vanity, literary flamboyance, and an excessive desire to appear clever by making the weaker side look stronger. He was deeply sorry for this transgression, and was ready to make amends.

This admission spared Galileo more drastic punishments, which might have ended with being burned alive at the stake. Such a capital punishment was meant to be charitable to the defendant, by giving him one last chance to repent, and thus save his soul, before his body was consumed by fire. However, Galileo declined such charity; he did *not* want to contribute to the "publish and perish" tradition, so to speak.

After this deposition, Galileo was allowed to return to the Tuscan embassy. On May 10, there was a third formal hearing at which he presented his defense. He repeated his recent admission of some wrongdoing, together with a denial of any malicious intent, and added a plea for clemency and pity. The trial might have ended here, but was not concluded for another six weeks. The new development was one of those things that make the Galileo affair such an unending source of controversy, and such rich material for tragedy.

Threat of Torture

Obviously, the pope and the cardinals of the Congregation of the Holy Office would have to approve the final disposition of the case. Indeed, it was standard Inquisition practice for an official, the assessor, to compile a summary of the proceedings for the benefit of the cardinal-inquisitors. So a report was written, summarizing the events from 1615 to Galileo's third deposition just completed.

Through a series of misrepresentations, this report left no doubt that Galileo had committed some kind of criminal act. On the other hand, by various quotations from the confessions and pleas in his depositions, the report made it clear that he was not obstinately incorrigible, but rather was sorry and willing to submit.

However, the report did not resolve the pope's and the Inquisition's doubts about Galileo's intention. So, on June 16, at an Inquisition meeting presided over by Pope Urban, they reached several decisions.[6] The defendant should undergo a so-called "rigorous examination"; that is, he should be interrogated under the verbal threat of torture in order to determine his intention. Even if his intention was found to have been untainted and pure, he had to recite an abjuration. He was to be condemned to formal arrest at the pleasure of the Inquisition, and the *Dialogue* banned.

Threat of torture and actual torture were, at the time, the standard practice of the Inquisition,[7] and indeed of almost all systems of criminal justice in the world. Nevertheless, such an interrogation, together with the abjuration, the formal imprisonment, and the book ban were not really in accordance with the spirit or the letter of the out-of-court plea-bargaining and agreement. Galileo felt betrayed and always remained bitter about this outcome.

On June 21, Galileo was subjected to the formal interrogation under the verbal threat of torture. He did not undergo actual physical torture, since the papal decision had been only to use the verbal threat, and the Inquisition was scrupulous about such distinctions. Moreover, the Inquisition had rules and practices about torture, which it usually followed. These rules prohibited actual torture from being applied to those who were elderly or in ill health; and Galileo was both.

The result of the interrogation was favorable. For, even under such a threat, Galileo denied any "malicious" intention—i.e., he denied that his defense of the Earth's motion had been intentional, in the sense of being motivated by a belief that the Earth moves; and he showed his readiness to undergo actual torture, and to die if need be, rather than admit to such an intention and belief.

Sentence

The following day, June 22, at a ceremony in the convent of Santa Maria sopra Minerva in Rome, he was read the sentence and then recited the formal abjuration.

The sentence[8] found Galileo "vehemently suspected of heresy." The expression "vehement suspicion of heresy" was, as we saw earlier, a technical legal term which meant much more than it may sound to modern ears. The Inquisition was primarily interested in two main categories of crimes: formal heresy and suspicion of heresy. One difference between them was the seriousness of the offense; another was whether the culprit, having confessed to the incriminating facts, admitted having an evil intention. Furthermore, two main subcategories of suspicion of heresy were distinguished, vehement and slight, again dependent on the seriousness of the crime.

"Suspicion of heresy" was not, then, merely suspicion of having committed a crime, but was itself a specific category of crime. Galileo was in effect being convicted of an offense that was intermediate in seriousness, among those handled by the Inquisition.

The sentence mentioned not one but two main errors, or suspected heresies. The first involved holding a doctrine that was false and contrary to Scripture. The second involved assuming the principle that it is permissible to defend as probable a doctrine contrary to Scripture.

This two-fold character of Galileo's suspected heresy is important. The first error concerned a doctrine about the location and behavior of natural bodies. Four particular theses were distinguished—heliocentrism, heliostaticism, geokineticism, and the denial of geocentrism; but the respective status of these four parts was not being censured in any more nuanced fashion. The second suspected heresy concerned a methodological, epistemological, theological, and hermeneutical principle, a norm about what is right or wrong in the search for truth and the quest for knowledge. It was a principle that declared Scripture to be irrelevant to physical investigation, and denied the authority of Scripture in natural philosophy (physics, astronomy, science).

Recall that these two suspected heresies were the main charges of which Galileo had been accused in 1615 by Caccini and Lorini. Thus, the sentence was presenting the condemnation of 1633 in part as a resumption of the proceedings temporarily suspended in 1616, and as confirmation of the essential correctness of those charges, now demonstrated by the publication of the *Dialogue*.

Furthermore, the methodological principle for which Galileo was being condemned was expressed in terms of probability: "that one may hold and defend as probable an opinion after it has been declared and defined contrary to the Holy Scripture."[9] This seems to add another level of nuance and prohibition to the ones we discussed earlier. Let me explain.

In our discussion of Galileo's first confrontation with the Inquisition in 1616, we distinguished several orders to him regarding Copernicanism. There was Pope Paul V's injunction *not to discuss* Copernicanism, intended to be issued in case Galileo would reject Bellarmine's warning; it is clear that this order was never actually issued by the commissary to Galileo. Then there was Commissary Seghizzi's precept *not to hold, defend, or teach in any way* the Copernican doctrine, which is the order described in the notary memorandum; the illegitimacy and likely non-occurrence of this order are also supported by the evolution of the 1633 proceedings, since Galileo denied having received it and the prosecution dropped the related charge in light of Bellarmine's certificate. Then there was Bellarmine's warning *not to hold or defend Copernicanism as true, or as compatible with Scripture, but only as a hypothesis*; this order was obviously issued, admitted as received by Galileo, and eventually confessed as unintentionally violated by him. However, before the 1633 sentence, there was no distinct order *not to hold or defend Copernicanism as probable*, so this may be regarded as the addition or invention of a new prohibition. Or, more charitably, perhaps this formulation in the sentence is essentially interpreting the term "true" in Bellarmine's warning to include "probably true." But to be charitable to Galileo, we could say that he too is sticking to Bellarmine's warning, but interpreting the term "hypothesis" to include "probable, although not yet demonstrated, truth."

In the text of the sentence, the verdict was followed by a list of penalties. The first was that Galileo was to immediately recite an "abjuration" of the "above mentioned errors and heresies." Second, the *Dialogue* was to be banned by a public edict. Third, Galileo would be kept under imprisonment indefinitely. And fourth, he would have to recite the seven penitential psalms once a week for three years. Finally, the Inquisition declared that it reserved the right to reduce or abrogate any or all of these penalties.

We will see presently what the abjuration amounted to. But note that, although it was a penalty in the sense that it was a great humiliation for anyone to have to recant one's views, it was also a procedural step for the culprit to gain absolution of the sin of heresy.

The book's prohibition took effect immediately, although a year passed before it would be included in a formal decree of the Congregation of the Index that listed other books as well. The recitation of the penitential psalms was a "spiritual" penance for the good of Galileo's soul.

The "imprisonment" stipulated in the sentence never did involve detention in a real prison, although it did last for the rest of his life. For two more days, Galileo was confined to the prosecutor's apartment at the Inquisition palace, which had been his place of detention for certain periods of the trial, April 12–30 and June 21–2. For about a week, June 24 to July 1, Galileo was under house arrest at Villa Medici, the sumptuous palace in Rome belonging to the Tuscan grand dukes. Then his place of detention was commuted to the residence of the archbishop of Siena, who was a good friend of Galileo's; he stayed there from early July to early December. Finally, from December 17, 1633 onward, Galileo was under house arrest at his own villa in Arcetri near Florence.

However, these details, which are available to us today, were generally unknown until the relevant documents were discovered and published at the end of the eighteenth century. So, for about 150 years, on the basis of the sentence one could justifiably conclude that Galileo had been put in prison as a result of his condemnation. This led to the growth of the prison myth,[10] which eventually acquired a life of its own.

The sentence ended with the signatures of seven out of the ten cardinal-inquisitors. This fact would hardly need comment were it not that three signatures are thus missing—those of Francesco Barberini, Gasparo Borgia, and Laudivio Zacchia. Francesco Barberini was the pope's nephew and the Vatican secretary of state; he was the most powerful man in Rome after the pope himself. Borgia was the Spanish ambassador and leader of the Spanish party in Rome; a year earlier, he had threatened the pope with impeachment on account of his behavior in the Thirty Years' War. Zacchia was the chief of staff of the papal household. Given the powerful positions of these three cardinals, the explanation of the missing signatures has become one of the many controversial questions in the Galileo affair. Two obvious possibilities are that the lack of these signatures reflected a significant disagreement among the ten judges, or merely that some of them were not present at the meeting when the document was signed.

Abjuration

The most immediate penalty for Galileo's crime and the final procedural step in the trial was for Galileo to recite an "abjuration."[11] Its content and wording were relatively standardized and provided to him by the Inquisition's officials.

The document begins with a multifaceted confession. Galileo admitted having been notified that the heliocentric, geokinetic doctrine was contrary to Scripture and so could not be held or defended, and having subsequently published a book that held and defended this doctrine. These two admissions corresponded to what had actually happened during the proceedings and to what the sentence had reported.

However, Galileo also admitted to having been served with the full special injunction, i.e., with the judicial precept to completely abandon the doctrine and not to hold, defend, or teach it in any way whatever, orally or in writing. This admission went beyond his previous confession and beyond the sentence, which had allowed that Bellarmine's certificate cast doubt on the special injunction having been served. So the text of the abjuration made Galileo confess additionally to violating the special injunction besides unintentionally violating Bellarmine's warning.

Galileo then acknowledged that because of these transgressions, he had been judged to be vehemently suspected of heresy, and that this verdict was right when he spoke of "this vehement suspicion, rightly conceived against me."[12] He was confessing that he was guilty as judged.

After these confessions, we come to the abjuration proper, where the culprit "with a sincere heart and unfeigned faith"[13] abjured and cursed the above-mentioned errors and heresies. This has led to the question whether such Galilean abjuring amounted to perjury—another classic controversy of the Galileo affair.

The culprit then made a series of solemn promises about future thought and behavior: that he would never again hold any such beliefs; that he would denounce to Church officials any heretics or suspected heretics; that he would comply with the penalties imposed on him by the judges; and that he would submit to further penalties if he failed to comply with the current ones. Such promises suggest something which was entirely normal for Inquisitorial practice: that at the final session of the trial Galileo read this document word for word and affixed his own handwritten signature to it at the end, but that the text had been compiled by the Inquisition's clerks.

Whether this abjuration was extorted from Galileo during the "rigorous examination," perhaps in exchange for doing without additional "rigors," as some scholars have claimed,[14] is a controversial question that we can leave here simply as an issue to be added to our accumulating stock. But it ought to be clear by now and it certainly is striking that in this abjuration Galileo was not only being made to retract earlier beliefs, attitudes, and behavior, but also being made to plead guilty to the verdict already announced by the judges and to confess to a transgression not confessed earlier.

Thus, the original Galileo affair ended and a new one began. What ended was the Inquisition's trial of Galileo, which started in 1613 with his letter to Castelli, refuting the biblical objection to the geokinetic hypothesis, and which climaxed in 1633 with his condemnation as a suspected heretic, for writing a book that defended this hypothesis and rejected the astronomical authority of Scripture. What began was the unresolved and perhaps unresolvable controversy about the facts, the issues, the causes, and the lessons of the original episode. This is a cause célèbre which continues to our own day and whose fascination rivals that of the original one.

CHAPTER 8

BECOMING A CULTURAL ICON (1616–2016)

The Subsequent Galileo Affair

The subsequent Galileo affair, the controversy about the original episode that started in 1633 and continues to this day, is much more complex than the original one, because of the longer historical span, the broader interdisciplinary relevance, the greater international and multilinguistic involvement, and the ongoing cultural import. To begin to make sense of it, we need to consider three aspects: the historical aftermath; the reflective commentary; and the critical issues.

By historical aftermath I mean the events directly stemming from the trial and condemnation of Galileo. Some of these involve actions taken by the Catholic Church, such as the partial unbanning of Galileo's *Dialogue* by Pope Benedict XIV in 1744, and most recently the rehabilitation of Galileo by Pope Saint John Paul II (in 1979–92). The historical aftermath also includes actions by various non-ecclesiastic actors, such as René Descartes's decision (in 1633) to abort the publication of his own cosmological treatise *The World*; and the attempts in the middle of the twentieth century by various secular-minded and left-leaning intellectuals, such as Bertolt Brecht, Arthur Koestler, and Paul Feyerabend, to blame Galileo for many current social and cultural problems.

The reflective commentary on the original trial consists of countless interpretations and evaluations advanced in the past four centuries by scientists, theologians, historians, philosophers, writers, cultural critics, and many others. These comments have appeared sometimes in specialized scholarly publications, sometimes in private correspondence or confidential ecclesiastical documents, and sometimes in classic texts. Among the latter are Descartes's *Discourse on Method* (1637), John Milton's *Areopagitica* (1644), Blaise Pascal's *Provincial Letters* (1657), Gottfried Leibniz's *New Essays on Human Understanding* (1704), Voltaire's *Age of Louis XIV* (1751), Denis Diderot and Jean D'Alembert's French *Encyclopedia* (1751–77), Auguste Comte's *Positive Philosophy* (1830–42), John Henry Newman's (1801–90) writings, Brecht's *Galileo* (1938–55), and Koestler's *Sleepwalkers* (1959). Here we have a historiographical or meta-historical labyrinth in which it is easy to get lost unless one uses some tentative guidelines; for example, by distinguishing accounts that are circumstantial from principled, one-dimensional from multifaceted, pro- from anti-Galilean, pro- from anti-clerical, and neutral from polemical.

The critical issues of the subsequent controversy in part reflect the original issues, which involved questions like whether the Earth is located at the center of the universe; whether the Earth spins around its own axis daily and orbits the Sun; whether and how the Earth's motion can be proved, experimentally or theoretically; whether the Earth's motion contradicts Scripture, and whether a contradiction between terrestrial motion and a literal interpretation of Scripture would constitute a valid reason against the Earth's motion; whether Scripture must indeed always be interpreted literally and, if not, when it should be interpreted figuratively. However, the subsequent controversy has also acquired a life of its own, with debates over new issues such as whether Galileo's condemnation was right; whether science and religion are incompatible; what useful lessons can be derived from the original episode about scientific methodology and human rationality in general; how science and religion do or should interact; whether individual freedom and institutional authority must always clash; whether cultural myths can ever be dispelled with documented facts; whether political expediency should prevail over scientific truth; and whether scientific research must bow to social responsibility.

Although distinct, these three principal aspects of the subsequent Galileo affair are obviously interrelated. For example, much of the reflective commentary consists of attempts to formulate or resolve one or more critical issues, and such formulations often represent important developments of the historical aftermath. Each moreover has substrands and finer subdivisions as well as broader groupings. For example, the ecclesiastical part of the historical aftermath includes at least two distinct clusters of developments: one is aimed to repeal the censures against the Copernican doctrine and books; the other strives to "rehabilitate" the person Galileo or apologize for his persecution. Similarly, the lay part can usefully be divided into several clusters, including a series of scientific discoveries and developments that demonstrate the reality of the Earth's motion, and the many post-trial and posthumous attempts by some to defend and re-affirm the Inquisition's condemnation, and by others to defend Galileo and criticize the Inquisition. In this chapter, I focus on some of the most interesting and consequential aspects of the subsequent Galileo affair, without pretending to give a complete account.

Unbanning Books

The two main books prohibited by the decree of the Index in 1616, at the end of the first phase of Galileo's trial, were Copernicus's *Revolutions* and Paolo Foscarini's *Letter on the Pythagorean Opinion*. Moreover, the decree explicitly prohibited all other books teaching the same doctrines.

Thus, in 1619, when a new supplement to the *Index of Prohibited Books* was published, it included these books, as well as an entry that read: "all books that teach the motion of the earth and the immobility of the sun."[1] This supplement also listed Kepler's *Epitome of Copernican Astronomy*, which had just been banned the same year. In 1633, the sentence against Galileo included the banning of his *Dialogue*, and so the book was included in a formal decree of the Index the following year.

The net effect of these prohibitions was that Catholics were not allowed to read such books unless they received special formal permission from the Church. With regard to holding geokinetic beliefs, the situation was less clear. The safe thing to do was to avoid the topic altogether, although one was allowed to engage in a so-called "hypothetical" discussion.

Eventually, these prohibitions were lifted. The unbanning process was gradual and slow, and took about two centuries. In 1620, as we have seen, the Index issued the corrections to Copernicus's book promised in 1616; with the stipulated passages deleted and others appropriately changed, the book could be read. In 1744, during the papacy of Benedict XIV, Galileo's *Dialogue* was republished for the first time with the Church's approval, as the fourth and last volume of his collected works; in the volume, the text of the *Dialogue* was preceded by the Inquisition's sentence and Galileo's abjuration of 1633 and by a supposedly neutralizing introduction. In 1757, upon request from Pope Benedict XIV, the Congregation of the Index dropped from the forthcoming edition of the *Index* the clause "all books that teach the motion of the earth and the immobility of the sun," and so the following year's edition no longer listed as an entry this general prohibition; but it continued to include the previously prohibited books by Copernicus, Foscarini, Kepler, and Galileo. In 1820, the Inquisition gave the imprimatur to an astronomy textbook by a professor at the University of Rome named Giuseppe Settele that presented the Earth's motion as a fact; in so doing, it overruled the objections of the chief censor in Rome, the so-called master of the sacred palace. Two years later, the Inquisition decided that in the future this official must not refuse the imprimatur to publications teaching the Earth's motion in accordance with modern astronomy; but a decision about removing from the *Index* the listed particular Copernican books was postponed. In 1833, while deliberating on a new proposed edition of the *Index*, Pope Gregory XVI decided that it would omit these books, but without explicit comment. So the 1835 edition of the *Index* for the first time omitted from the list Copernicus's *Revolutions*, Galileo's *Dialogue*, and the other books. This was the final and complete retraction of the book censorship begun in 1616.

The *Index* itself continued a while longer. In 1912, Pope Benedict XV abolished the Congregation of the Index as a separate department of the Church, and he made book censorship one of the tasks of the Congregation of the Holy Office, that is, the Inquisition. In 1966, the very institution of a list of prohibited books was abolished when the supreme Congregation for the Doctrine of the Faith (the new name of the Inquisition) decreed that the book censorship now under its jurisdiction no longer has the force of ecclesiastical law, but only the status of moral advice.[2]

Rehabilitating a Heretic

With regard to the condemnation of Galileo the individual, the history of its ecclesiastic aftermath is more elusive, complex, and controversial.[3] The climax, though not the conclusion, of this strand of the story is an episode that unfolded at the end of the twentieth century, having begun in 1979 and having had a formal conclusion in 1992. This is what some have called the rehabilitation of Galileo by Pope Saint John Paul II. However, others have described it as an attempt at a self-rehabilitation by the Church based on Galileo; and still others regard it as a new myth about Galileo—the myth that the Church has rehabilitated Galileo.

The alleged rehabilitation began very promisingly in 1979, when the pope included many pro-Galilean comments in a speech to the Pontifical Academy of Sciences, occasioned by the celebration of the centennial of Albert Einstein's birth. This was followed in 1980 by an announcement of the creation of a Vatican Commission on Galileo. Then, for the next several years, the Pontifical Academy of Sciences and the Vatican Astronomical Observatory sponsored conferences and publications meant to improve the documentation of the facts of the original trial and to clarify the issues. The process was formally concluded in 1992 at a meeting of the Pontifical Academy, when the pope made a speech accepting a report by Cardinal Paul Poupard, the head of the Galileo Commission.

The upshot, in my view, was as follows. In his two speeches to the Pontifical Academy of Sciences, and other statements and actions, John Paul admitted that Galileo's trial was not merely an error, but also an injustice. Moreover, the pope was clear and explicit that Galileo was theologically right about scriptural interpretation, as against his ecclesiastical opponents. Furthermore, the pontiff added an important point concerning the pastoral aspect of the affair: pastorally speaking, Galileo's desire to disseminate novelties was as reasonable as his opponents' inclination to resist them. Finally, according to the pope, Galileo provides an instructive example of the harmony between science and religion; and the example involves both words and deeds, in the sense that he both lived such harmony and justified it with good arguments.

This rehabilitation was informal because the pope was merely expressing his personal opinions and not speaking *ex cathedra*. It was also partial because he deliberately avoided talk or action regarding a formal judicial retraction or revision of the 1633 Inquisition trial and sentence. Furthermore, the rehabilitation was muddled; it was opposed by various elements of the Church, including some in the Vatican Commission on Galileo, such as Cardinal Poupard himself. They attempted to repeat many of the traditional apologias, for example, the account elaborated by Pierre Duhem, who, as we shall see, claimed that Church officials understood better than Galileo the limitations of his pro-Copernican arguments and of scientific argumentation in general. Nevertheless, the rehabilitation was significant, indeed epoch-making; and in this regard, it should be mentioned that John Paul was the first Polish pope since Copernicus, and the first non-Italian pope since Galileo.

One final point about John Paul's rehabilitation of Galileo: it was not unprecedented. Indeed, it was the sort of process one would have expected in light of precursors in the previous four centuries of the Galileo affair. One of these episodes is still unfolding, but pertains to a development going back more than half a century. The occasion was provided by the 1942 tricentennial of Galileo's death. In fact, the tricentennial engendered a first partial, informal, and muddled rehabilitation.

In the period 1941–6, various pro-Galilean pronouncements were expressed by several clergymen who held the top positions at the Pontifical Academy of Sciences, the Catholic University of Milan, the Pontifical Lateran University in Rome, and the Vatican Radio. They published accounts of Galileo as a Catholic hero who upheld the harmony between science and religion; who had the courage to advocate the truth in astronomy even against the Catholic authorities of his time; and who had the religious piety to retract his views outwardly when the 1633 trial proceedings made his obedience necessary.

But there is another, darker, side to this episode. In 1941, the Pontifical Academy of Sciences commissioned Monsignor Pio Paschini to write a book on Galileo's life and work and their historical background and significance. Paschini completed his manuscript in 1944 and submitted it for approval. During the following year, various ecclesiastic authorities (the Pontifical Academy of Sciences, the Vatican Astronomical Observatory, and the Holy Office) judged Paschini's manuscript to be unsuitable for publication, on the grounds that it was too favorable to Galileo and too critical of the Jesuits and the Church. After various initial appeals, Paschini abandoned hope of having his work published, and remained silent about it for the rest of his life.

Paschini died in 1962, and then his legal heir undertook a successful effort to have the manuscript published. At this time, the Church favored publishing the book, partly to celebrate the four-hundredth anniversary of Galileo's birth in 1964, and so the book was published that year by the Pontifical Academy of Sciences. In part, the Church wanted to have an intellectual foundation for some of the deliberations at the Second Vatican Council; in fact, in some of the published documents of that Council, there are footnote references to Paschini's book.[4]

However, in 1979–80, at about the same time that Pope John Paul was starting his rehabilitation, the original manuscript of Paschini's book began to be examined by various scholars, and to be compared to the published book. They discovered that the published version contains so many and such significant emendations that it must be regarded as an adulteration of the original. In fact, in 1964, when the Church re-examined the question of publishing Paschini's book, the Pontifical Academy of Sciences had charged Jesuit father Edmond Lamalle with reviewing and revising the manuscript. The published book is a Jesuit version of Paschini's manuscript.

Ironically, it is Paschini's published book that is most frequently referred to in Pope John Paul's various speeches and essays. And the unadulterated version of the book has not been published to date.

So, intriguingly, in the period 1979–92, two developments were occurring simultaneously: the alleged rehabilitation of Galileo by Pope John Paul and the second phase of the Paschini episode. An uncharitable interpretation would be that in the past four centuries the Galileo affair has undergone a metamorphosis: from the censoring of science books to the censoring of history-of-science books.

Another important milestone was also a kind of precursor of John Paul's rehabilitation, but antedates it by about a century. In 1893, Pope Leo XIII published an encyclical letter entitled *Providentissimus Deus*. This document put forth a view of the relationship between biblical interpretation and scientific investigation that corresponded to the one advanced by Galileo in his letters to Castelli and Christina. To be sure, the encyclical did not mention Galileo, but it was written in response to the controversy known as "the biblical question." This was concerned with the nature, methods, and implications of the scientific study of the Bible and the validity of the scientific criticism of Scripture.

The problem of the scientific criticism of Scripture was the reverse of what Galileo had to deal with: he was trying to defend astronomical theory from objections based on scriptural assertions, whereas Pope Leo was discussing how to defend Scripture from attempts to criticize its content based on physical science. But their respective answers hinged on essentially the same point: the denial of the scientific authority of Scripture. Not only were both Galileo and Leo asserting the same principle, they also shared crucial aspects of the reasoning to justify this principle.

In fact, one of Leo's key arguments should ring familiar in light of our earlier discussion of Galileo's *Letter to Christina*: he claimed that natural science and scriptural assertions cannot contradict each other, because both nature and Scripture derive from God. So if there appears to be a contradiction, the conflict is only apparent and not real, and must be resolved. This is normally done by interpreting the biblical statement in a nonliteral fashion, which means we have to say that Scripture is not a scientific authority. Given this, the principle of accommodation also follows.

Besides the formal similarity of problems, the substantive overlap of content, and the deep correspondence of the reasoning, Leo's account was reminiscent of Galileo's even in its appearance, in quotations from St. Augustine and how they were interwoven with the rest of the argument. In fact, Leo's two main passages from Augustine had also been quoted by Galileo in his *Letter to Christina*: Augustine's statement of the priority of demonstrated physical truth ("whatever they can really demonstrate to be true of physical nature, we must show to be capable of reconciliation with our Scripture"[5]) and his statement of nonscientific authority of Scripture ("the Holy Ghost…did not intend to teach men…the things of the visible universe"[6]).

It's not surprising that Leo's encyclical has been widely perceived to have been the Church's belated endorsement of the second fundamental belief for which Galileo had originally been condemned, namely that Scripture is not an authority in astronomy. And besides being belated, the endorsement was silent and merely implicit. However, as we have seen, on this issue, Pope John Paul II made such a vindication of Galileo explicit.

The history of this strand of the subsequent Galileo affair is by no means limited to recent events. In fact, this aspect of the story began immediately after his death, when questions were raised about whether a convicted heretic like him had the legal right to have his will executed and to be buried on consecrated ground. These issues were decided in his favor, after they were investigated by lawyers and some formal legal opinions were formulated. However, the same did not happen regarding the question whether it was proper to build an honorific mausoleum for Galileo in the Church of Santa Croce in Florence. This was initially vetoed by the Church in 1642, but it was brought about a century later in 1737. Since that time, Galileo's body has been buried in such a mausoleum across from Michelangelo's tomb (see Figure 19).

Finally, we should mention the story of the publication of the special Vatican file containing the manuscripts of the original trial documents.

Figure 19. Galileo's tomb, at the Church of Santa Croce, Florence

In 1780, an inquiry was made by Giovanni Targioni Tozzetti, a pioneer in the history of science, who was the first to publish many important documents of seventeenth-century Italian science. He reported with disappointment, having been told by Inquisition officials, that there were no documents pertaining to Galileo's trial in the Archives of the Holy Office. This was indeed true, for the trial documents were not kept in the relatively obvious place where Tozzetti inquired, but rather in the so-called Vatican Secret Archives, befitting their special status and significance. This was discovered in 1810, when, by order of the French emperor Napoleon, the file was taken to Paris, as part of his decision to transfer to France all archives of the Vatican, the Inquisition, and other Church congregations in Rome. For the next seven years, the Church tried unsuccessfully to retrieve the file before giving up, having concluded that it had been lost or destroyed. It finally re-surfaced in Vienna in 1843 and was promptly delivered back to Rome.

This kind of attention received by the file soon led to its publication by lay scholars who received the Church's permission to consult it. A partial edition was first published in 1867 by the French scholar Henri de L'Epinois. Then the complete dossier was published in 1876 by the Italian Domenico Berti, and in 1877 by the Austrian Karl von Gebler. These Vatican documents and many others were then included in the National Edition of Galileo's collected works published in 1890–1909. As a result, the discussion of the affair moved to a qualitatively higher plane.

The publication of the trial proceedings in the 1860s and 1870s led to a new controversy and eventually to a consensus concerning a previously inacccessible aspect of the trial. This was the controversy concerning the authenticity, accuracy, and legal validity of Commissary Seghizzi's injunction to Galileo, not to hold, defend, or teach the Earth's motion in any way whatever. A common scholarly opinion at that time was that the document is a forgery, and hence factually questionable and legally invalid. But as we have seen, the most tenable position is that, although the document is authentic and not a forgery, the injunction was certainly illegitimate, and the report is probably inaccurate.

Proving the Earth's Motion

The key scientific claim for which Galileo was tried and condemned, the proposition that the Earth revolves on its axis and orbits the Sun,[7] inflamed a scientific controversy. It was a controversy which had existed since

Copernicus's *Revolutions* of 1543, but it now took a more intense and more definite form: more definite, because it now focused on whether the Earth moves and whether this motion can be proved or disproved experimentally by either terrestrial or astronomical evidence; and more intense, because scores of books were published, new experiments devised and performed, new arguments invented, and old arguments re-hashed.

The most important of these developments happened in 1687, when Isaac Newton published his *Mathematical Principles of Natural Philosophy* (the *Principia*). The Newtonian system of celestial mechanics has two important geokinetic consequences, among others. First, the relative motion between the Earth and the Sun corresponds to the actual motion of both bodies around their common center of mass; but the relative masses of the Sun and the Earth are such that the center of mass of this two-body system is a point inside the Sun; so, although both bodies are moving around that point, the Earth is circling the body of the Sun. Second, the daily axial rotation of the Earth has the centrifugal effect that terrestrial bodies weigh less at lower latitudes, and least at the equator, and that the whole Earth is slightly bulged at the equator and slightly flattened at the poles; and these consequences were verified by observation. In other words, these observational facts can be explained in no other way than by terrestrial rotation.

However, these Newtonian proofs were still relatively indirect, and so the search for more direct evidence continued. In 1729, English astronomer James Bradley discovered the aberration of starlight, providing direct observational evidence that the Earth has translational motion in space. In 1806, Giuseppe Calandrelli, director of the astronomical observatory at the Jesuit Roman College, claimed to have measured the annual parallax of the brightest star in the constellation Lyra; the variation was about 5 seconds of arc, yielding a distance of about 250 light days, which is a value about 15 times smaller than the actual 11 light years. Thus, the discovery of annual stellar parallax is usually attributed to German astronomer and mathematician Friedrich Bessel, who observed it for the star 61 Cygni in 1838. Annual stellar parallax provides direct proof that the Earth revolves annually in a closed orbit. In 1851, Léon Foucault in Paris invented the pendulum that bears his name and provided a spectacular demonstration of the Earth's rotation; the experiment was ceremoniously repeated in many other places. Finally, in 1910–11, Jesuit J. G. Hagen, at the Vatican Astronomical Observatory in Rome, invented a new instrument (called isotomeograph) to demonstrate and measure the Earth's rotation in a new way.

Of course, by now we are relatively far removed from the Galileo affair or even the motion of the Earth. The challenge in these latter experiments is not to prove or confirm something which is uncertain, but rather to measure with a very high level of precision a phenomenon which is uncontroversial. There are various technical challenges, as physicists now try to take into account such things as the detailed shape of the Earth, which is not perfectly spherical, the density variations within falling bodies, and even the effect of lunar gravitational attraction.

To bring us back to the Galileo affair and the Earth's motion, let's consider the contribution of an Italian priest and mathematician named Giambattista Guglielmini, in 1789–92. It was he who pioneered the detection of the eastward deviation of falling bodies, thus confirming eastward terrestrial rotation.

Recall the Aristotelian objection to terrestrial rotation based on vertical fall: if the Earth were rotating eastward, bodies in free fall would be left behind, and so would exhibit a trajectory slanted westward, landing at a spot on the ground west of the point of release; and since observation reveals that bodies fall vertically, it follows that the Earth does not rotate. In the *Dialogue*, Galileo answered this objection in great detail, as we will see. Here, I will simply mention one point in his answer: he argued that if the Earth were rotating, the horizontal motion which a body would have before being released would be conserved, and it would combine with the downward motion during the fall to bring the body exactly under the point of release without deviation, so that we would observe the same vertical fall as we would on a motionless Earth.

This reply is correct as a first approximation. But Galileo himself gives various hints elsewhere in the book to the effect that, on a rotating Earth, falling bodies would actually *advance* forward horizontally as they fall, and so would be deflected *eastward*.[8] The basic reason for this is that on a rotating Earth a body at the top of a tower is moving horizontally with a faster linear speed than the base of the tower, due to the fact that the geocentric circumference at the top of the tower is longer than that at the base by an amount which is a function of the height of the tower. Following up on such hints, Guglielmini computed the amount of such an eastward deviation, devised some experiments to detect it, and confirmed the predictions. The predicted deviation is of course very small; for a height of about 160 feet used by Guglielmini, the deviation was calculated to be about 2/3 of an inch. Thus, all kinds of precautions had to be taken and devised, involving such things as the mechanism of release; stopping microscopic pendular vibrations before release; minimizing disturbances from outdoor traffic, winds, and temperature; and averaging out the spread of deflections from one trial to another.

In 1804, the eastward deviation was also confirmed in Germany by Johann Benzenburg. In the meantime, more sophisticated and precise calculations were worked out by Pierre-Simon Laplace (1749–1827) in France and Carl Friedrich Gauss (1777–1855) in Germany, working independently of each other and using different methods and principles; they predicted an eastward deviation with a value about 2/3 of that calculated by Guglielmini. As late as the first decade of the twentieth century, the phenomenon continued to attract some attention; in 1903–10, physicist Edwin Hall, from the Jefferson Laboratory at Harvard University, gave an updated sophisticated experimental confirmation of Guglielmini's eastward deviation of falling bodies and also of a predicted negligible southward deviation.

Finally, it is intriguing to note that when Guglielmini first conceived his experiments, he happened to be in Rome, attached to the entourage of a powerful cardinal. His idea was to exploit the greatest height of fall available at the time, which happened to be the dome of St. Peter's church. So he dreamed initially of providing an experimental proof that Galileo was scientifically right by dropping balls from the ceiling of that same church which had been the physical focus of his troubles.

Retrying Galileo

There were also various attempts after the trial to criticize Galileo and defend the Church, and these form the subject of much of the reflective commentary.[9] The anti-Galilean critiques and the pro-clerical apologias involve various points of view—scientific, logical, philosophical, theological, legal, moral, etc. Here, we will deal primarily with the main factual details of this story, and I will come to a critical interpretation of it later.

After Galileo's condemnation in 1633, it was only natural to want to know why he was condemned. It was equally natural to consider whether he had been unjustly persecuted by the Inquisition, and this varies depending on whether one is taking the point of view of science, philosophy, theology, law, morality, or practical utility. The first question pertains to the interpretation of the condemnation, the second to its evaluation. These two interrelated issues have been persistent themes of the subsequent Galileo affair.

One initial response by critics of Galileo and pro-clerical thinkers was to hope or try to show that he had been scientifically wrong. For example, in 1642–8, a controversy developed regarding the correctness of his science of motion; a controversy that has been called "the Galilean *affaire* of the laws of motion... a second 'trial'."[10] It started in 1642, when Pierre Gassendi elaborated the connection between the new Galilean physics of motion and Copernican astronomy, strengthening each. The focal point was Marin Mersenne, whose correspondence and contacts facilitated and encouraged discussion. The result of this first "retrial" was a vindication of Galileo, who, ironically, had died the year the controversy began.

Similarly, in 1651, Jesuit Giovanni Battista Riccioli published a monumental work claiming that the Inquisition had been right and wise in condemning Galileo, both scientifically and theologically. Scientifically speaking, Riccioli argued that this was so chiefly because neither the Ptolemaic nor the Copernican, but rather the Tychonic, system was the correct one, so Galileo was wrong in holding that the Earth moves. Riccioli made a comprehensive examination of all the arguments to support his scientific choice. He even invented a new geostatic argument based on Galilean ideas, a Galilean argument against Galileo, so to speak. Although this particular argument was originally ignored, it was widely criticized after 1665, when it was republished in another book by Riccioli. The controversy spawned at least nine books, before subsiding four years later.[11] The consensus was that Riccioli was scientifically wrong, insofar as he was misunderstanding Galileo's physics. Even Riccioli himself seemed to admit scientific defeat. Once again, the objections of Galileo's scientific critics backfired against them, and they ended up being discredited, and he vindicated.

Eventually, as we have seen, the Earth's motion became an incontrovertible fact, conclusively proved beyond any reasonable doubt. However, long before that happened, as it was becoming clearer that Galileo had been right in believing that the Earth moves, another genre of anti-Galilean criticism and apologia of the Inquisition had been emerging. He began to be charged with believing what turned out to be true for the wrong reasons, on the basis of flawed arguments, or with the support of inadequate evidence.

Even during Galileo's own lifetime, his geokinetic argument from the tides had seemed not completely convincing. After Newton's correct explanation of the tides as caused by the gravitational attraction of the Moon (and also of the Sun), one could also claim that there was definitely an error in Galileo's theory that the tides were caused by the Earth's motion. And so the anti-Galilean critics started to mention the tidal argument as one of Galileo's bad reasons for believing what turned out to be true. Today this criticism continues to be one of the most common charges against Galileo.

In 1841, an anonymous article in a German journal inaugurated this kind of apologia in an explicit manner. It argued that the Inquisition rendered a service to science by condemning the Copernican theory when it had not yet been demonstrated to be true, and by condemning Galileo for supporting it with scientifically incorrect arguments. This anonymous article was originally attributed by some to a certain professor Clemens of the University of Bonn, but it was later shown (in 1878) by Karl von Gebler to be a translation of an Italian essay authored by Maurizio B. Olivieri, commissary of the Roman Inquisition and former general of the Dominicans.

The original essay circulated widely in manuscript form and was published only posthumously in 1872. Olivieri claimed that the mechanical objections to the Earth's motion depended crucially on the assumption that air has no weight; that therefore they could not be answered until the discovery that air has weight; that Galileo was not aware of this fact; and that the discovery was made after his death by Evangelista Torricelli and Blaise Pascal.

As I said earlier, there is more to being right than that one's beliefs and conclusions happen to be true, i.e., correspond to reality. It is also important that one's motivating reasons and supporting arguments be right. One's reasoning is at least as important as the substantive content of one's beliefs.

However, most such anti-Galilean charges are misapplied and can be refuted. Galileo's reasoning can be successfully defended; indeed, as I will argue later, it can be shown to be a model of critical thinking. For example, the just-mentioned criticism by Olivieri is historically untenable since Galileo was clearly aware that air has weight; this can be seen from Galileo's letters to Giovanni Battista Baliani of 1614 and 1630, and from a discussion in Galileo's 1638 book, *Two New Sciences*. The criticism was also scientifically misconceived because most of the mechanical difficulties depended on such questions as conservation and composition of motion and the principles of relativity and inertia, and not on the weight of air.

In any case, other issues were bound to arise, and did arise, in the process of coming to terms with the condemnation of Galileo. As we have seen, he got into trouble with the Church, and was formally condemned, in part for holding the principle that Scripture is not a scientific authority, that Scripture is irrelevant to the assessment of provable or probable claims about nature. This methodological and theological principle is much more elusive and controversial than the astronomical and scientific claim that the Earth moves, and the corresponding issues are more complex.

At first, some anti-Galilean critics mentioned this principle as one of Galileo's main errors. For example, in 1651 Riccioli, besides criticizing the geokinetic theory scientifically, elaborated a very conservative version of biblical fundamentalism, according to which the literal meaning of biblical statements must be held to be physically true and scientifically correct;

thus, allegedly, the Inquisition had been theologically wise in upholding the fundamentalist view against Galileo.

Again, however, eventually it turned out that Galileo was essentially right regarding this principle as well. As we have seen, a crucial episode in the theological vindication of Galileo came in 1893, with Pope Leo XIII's encyclical *Providentissimus Deus*. About a century later, Leo's implicit vindication was made explicit in Pope Saint John Paul II's rehabilitation of Galileo. In his 1992 speech, John Paul stated that "the new science, with its methods and the freedom of research that they implied, obliged theologians to examine their own criteria of scriptural interpretation. Most of them did not know how to do so. Paradoxically, Galileo, a sincere believer, showed himself to be more perceptive in this regard than the theologians who opposed him."[12]

However, once again, as it became increasingly clear that Galileo's theological principle was correct, his critics started to emphasize the reasons and arguments he had given to justify it. They tried to find all sorts of incoherences and inconsistencies in it: that his *Letter to Christina* contains not only assertions denying the scientific authority of Scripture, but also assertions affirming it; that on the one hand he objects to the use of biblical passages against his own astronomical claims, but on the other hand he tries to interpret the passage in Joshua 10:12–13 in geokinetic terms (i.e., he wants to have it both ways); that he tries to illegitimately shift the burden of proof by pretending that he is not obliged to prove Copernicanism, but rather theologians are obliged to disprove it; and that he wants both to appeal to the theological tradition (for example, by frequent quotations from St. Augustine) and to overturn it by a radically new principle.

However, Galileo can be defended from this criticism of his reasoning, for the criticism is itself criticizable as invalid. In fact, in this case its invalidity should be obvious from my earlier account of the *Letter to Christina*.

Still, the greater complexity of the scriptural issue created new possibilities for anti-Galilean criticism. Independently of the truth or falsity of the principle denying the scientific authority of Scripture, and independently of the validity or invalidity of Galileo's reasoning to justify it, he is sometimes criticized for his theological intrusion. It is objected that Galileo was not a professional theologian, and so he had no right to interfere with, or get involved in, exegetical and hermeneutical discussions. One reply is that Galileo *did* stay away from theological discussions until his scientific ideas were attacked on scriptural grounds; after that, he had every right to defend himself by refuting those attacks as fully as he did.

Another common criticism of Galileo charges him with pastoral imprudence. Many Catholic authors have argued that, although Galileo may have been right in astronomy and biblical hermeneutics, he was definitely wrong from the pastoral point of view; this requires that the mass of believers not be scandalized or misled by new discoveries, and so the dissemination of truth (if not its pursuit) must be careful not to upset popular beliefs too suddenly and must be mindful of the social and practical consequences of truth. To answer such criticism, I have little hesitation in appealing to Pope Saint John Paul II. As we saw earlier, one aspect of his rehabilitation of Galileo dealt precisely with this pastoral issue. Instead of siding with Galileo's opponents, John Paul's opinion was "that the pastor ought to show a genuine boldness, avoiding the double trap of a hesitant attitude and of hasty judgment, both of which can cause considerable harm."[13] He was not reversing the traditional anti-Galilean criticism, but rather he was denying it, and pointing out that the correct pastoral position is one of arriving at a judicious mean between the two extremes of too much conservation and too much innovation. So while he was not really siding with Galileo on the historical substantive issue, his rejection of the opposite side was in this context a pro-Galilean position.

More strangely, at one point in the subsequent affair, Galileo was blamed for holding and doing the *opposite* of what he actually held and did: it was alleged that he preached and practiced the principle that biblical passages be used to confirm astronomical theories. This criticism got started in 1784–5 with an apologia of the Inquisition by Jacques Mallet du Pan in a French magazine and with the printing of an apocryphal letter attributed to Galileo (but forged by Onorato Gaetani) in Girolamo Tiraboschi's *History of Italian Literature*. The view proved to be long-lasting and widely accepted for more than a century. It became a slogan, that "Galileo was persecuted not at all insofar as he was a good astronomer, but insofar as he was a bad theologian."[14] The myth seems to have acted as a catalyst in encouraging the proliferation of pro-clerical accounts and the articulation of pro-Galilean ones, making the discussion of Galileo's trial the cause célèbre it is today.

Another strand of anti-Galilean criticism focused on his alleged legal or judicial culpability. It claimed that the trial did not really deal with the astronomical and biblical issues just discussed. Galileo was condemned neither for being a good astronomer nor for being a bad theologian, but rather for something else—disobedience or insubordination. His crime was the violation of the ecclesiastical admonition which he received in February 1616. Admittedly, it is uncertain whether this admonition amounted simply to Cardinal Bellarmine's warning not to hold or defend the Earth's motion as true or biblically compatible, or to Commissary Seghizzi's special injunction not to hold, defend, or teach the theory in any way whatever. But in either case, Galileo's *Dialogue* violated the admonition. In this view, the violation of the special injunction is clear, direct, and incontrovertible. And a violation of Bellarmine's warning can be claimed to have occurred because the book does defend the truth of the Earth's motion by criticizing all arguments against it and advocating some arguments in favor.

Whether valid or invalid, this criticism cannot be summarily dismissed. It can be dated as far back as Tiraboschi's apologia in 1793, and it continues to be repeated, refined, and embellished. In my opinion, however, such criticism is untenable.

When it comes to Commissary Seghizzi's special injunction, as we have seen, it turned out that the relevant defense is contained in the documents and manuscripts that make up the special Vatican file of Galilean trial proceedings, which became available to scholars in the decade 1867–78. A consensus soon emerged that the special-injunction document has enough irregularities that this aspect of the proceedings must be regarded as embodying a legal or judicial impropriety. It isn't necessary to question the authenticity or factual accuracy of Seghizzi's injunction. It suffices to question its legal validity. From this point of view, the legal criticism of

Galileo again backfired against the critics. It emerged that Galileo had been the victim of an injustice in a way that had been previously unsuspected.

There remains, of course, the criticism that Galileo violated Bellarmine's milder warning. A possible answer to this criticism relates to what I said earlier about the interpretation and the evaluation of Galileo's *Dialogue*. The *Dialogue* discusses the Earth's motion by examining all the (non-theological) arguments on both sides; the examination includes not only a presentation and an analysis of the arguments, but also their assessment. Galileo was indeed taking the liberty of *evaluating* the arguments; but he was hoping that if he carried out the evaluation fairly and validly, his having engaged in the assessment of the arguments would not be held against him. He was taking the gamble that a correct assessment of arguments would not be seen as an objectionable defense of Copernicanism.

Although such a defense of Galileo has never, to my knowledge, been fully articulated and justified, traces of it can be found in the historical record. In February 1633, soon after Galileo reached Rome to stand trial, Francesco Niccolini (the Tuscan ambassador) had a meeting with Cardinal Francesco Barberini (the Vatican secretary of state and a member-judge of the Inquisition tribunal) to discuss the forthcoming proceedings. To the cardinal's charge that Galileo's *Dialogue* amounted to "reporting much more validly what favors the side of the earth's motion than what can be adduced for the other side," the ambassador replied that "perhaps the nature of the situation indicated this, and therefore he was not to blame."[15] In 1635, Nicholas Claude Fabri de Peiresc (1580–1637) hinted at it when he interpreted the *Dialogue* as a "philosophical play," by which he meant a problem-oriented discussion of the arguments, evidence, and reasons on both sides. And in 1943, Pio Paschini explicitly formulated such a defense of Galileo by stating that "it was not his fault if the arguments for the heliocentric system turned out to be more convincing."[16]

Moreover, with regard to Galileo's alleged disobedience of Bellarmine's warning, was the disobedience really illegitimate, or conversely, was the warning really legitimate? There is no conclusive argument justifying the legitimacy of the warning. It may have been one of the many abuses of power in this story. If the warning was not legitimate, then Galileo disobeyed an illegal order. And even if the warning was proper from the point of view of canon law, we may ask whether it was also proper from the *moral* point of view. Again, at worst Galileo may have committed a legal "misdemeanor" while in pursuit of a morally desirable aim, or while exercising a basic human or civil right.

One might think that the implicit theological vindication of Galileo by an influential pope in 1893, coming soon after his judicial rehabilitation by the meticulous scholarship of the 1870s, on top of the older and more gradual scientific vindication provided by the proofs of the Earth's motion climaxing with Foucault's pendulum (1851), would prevent or discourage further indictments or re-trials of the victim. However, to think so would be to underestimate the power of human ingenuity or the unique complexity of the Galileo affair. In fact, a novel apologia was soon devised by a great scholar who combined knowledge of physics, history, and philosophy—Pierre Duhem (1861–1916). In 1908 he advanced the new charge that Galileo was a bad epistemologist.

The criticism of Galileo as a bad epistemologist is often confused with, and is indeed related to, the criticism that he was a bad arguer or reasoner, that he did a poor job in proving the Earth's motion, in defending Copernicanism. But the two criticisms are distinct. The epistemological criticism of Galileo attributes to him wrong or untenable epistemological principles and practices, and then it connects such epistemological errors or naïveté with the tragedy of the trial.

The epistemological doctrine which Duhem found especially objectionable is "realism": it states that science aims at the truth about the world, and scientific theories are descriptions of physical reality that are true, probably true, or potentially true. Duhem was an advocate of epistemological "instrumentalism," according to which scientific theories are merely instruments for making mathematical calculations and observational predictions, and not descriptions of reality, and so they are not the sort of things that can be true or false, but only more or less convenient. Duhem tried to blame Galileo's trial on epistemological realism, allegedly shared by

Galileo and his Inquisition persecutors, and also on their failure to appreciate instrumentalism, which in that historical context was being allegedly advocated by Cardinal Bellarmine and Pope Urban VIII. Duhem's view was "that logic was on the side of Osiander, Bellarmine, and Urban VIII, and not on the side of Kepler and Galileo; that the former had understood the exact import of the experimental method; and that, in this regard, the latter were mistaken."[17] Here, Duhem's word "logic" should be taken to mean primarily "epistemology," and not reasoning.

Duhem's epistemological criticism of Galileo is important and influential. Nevertheless, it is untenable, primarily because under the heading of Galilean realism Duhem subsumes too many other epistemological principles besides the ideal of truth and description of reality; but these other attributions are conceptually arbitrary and textually inaccurate.

Another example of this genre of criticism may be gleaned from the work of more recent scholars. It claims that Galileo subscribed to the traditional Aristotelian ideal of science as strictly demonstrative, which he was never able to give up despite some flirtations with fallibilism or probabilism; that he believed he had provided a strict demonstration of the Earth's motion (with arguments such as his explanation of the tides); that much opposition to him was an attempt to make him aware of the nondemonstrative status of his arguments or the nonviability of the demonstrative ideal; and that the operative role of this problem is visible in such documents as Bellarmine's letter to Foscarini (1615), Galileo's "Discourse on the Tides" (1616), the *Dialogue* (1632), and the consultants' reports on the latter book produced during the 1633 proceedings. Such an account could be labeled the criticism of Galileo as a failed Aristotelian, or a failed demonstrativist.

This criticism overlaps with both Duhem's epistemological criticism and the criticism of Galileo's reasoning mentioned earlier. For in part this criticism faults Galileo's epistemological doctrine of demonstration or his epistemological awareness of the nature of his own geokinetic arguments; and in part this criticism impugns the reasoning used by Galileo to arrive at or to justify his geokinetic beliefs. However, such criticism can be rebutted, and it emerges that rather than being a failed Aristotelian demonstrativist,

Galileo is someone who was able to assimilate and transcend the Aristotelian ideal of science as demonstration.[18]

A measure of Duhem's influence is that it has spawned a genre of anti-Galilean criticism in which Galileo is charged with having held all kinds of implausible epistemological doctrines, and then a questionable connection with the trial is made. For example, a recent popular book revealingly entitled *Galileo's Mistake*[19] portrays Galileo as a kind of positivist who held that only science provides the truth about reality, and that this mistake was the root cause of his condemnation.

Current Cultural Developments

The subsequent Galileo affair shows no signs of abating to date. This is obvious not only from the recent rehabilitation efforts by the Catholic Church, but also from the recent anti-Galilean critiques by left-leaning social critics.[20]

At about the same time that Galileo was being rehabilitated by various Catholic officials and institutions, he became the target of unprecedented criticism from various representatives of *secular culture*. It was almost as if a reversal of roles was occurring, with his erstwhile enemies turning into friends, and his former friends becoming enemies.

These critics articulated what may be called social and cultural criticism of Galileo; they tried to blame Galileo by holding him personally or emblematically responsible for such things as the abuses of the Industrial Revolution, the social irresponsibility of scientists, the atomic bomb, and the gap between the two cultures (science vs. humanities). They were mostly writers with backgrounds and sympathies on the political left. The most outstanding and original examples of such criticism came from central-European German-speaking writers. The German playwright Bertolt Brecht's work *Galileo* was first written in 1938, then revised into a second version in 1947, and a third in 1955, and became a classic of twentieth-century theater. The Hungarian-born writer, novelist, and intellectual Arthur Koestler published *The Sleepwalkers: A History of Man's Changing Vision of the Universe* in 1959, and it became an international bestseller. And Austrian-born philosopher Paul Feyerabend advanced his version of social criticism in the book *Against Method*, first published in 1975 and revised in 1988 and 1993.

These developments have not yet been properly assimilated. For example, the Catholic "rehabilitations" tend to be either unfairly criticized (even by Catholics) or uncritically accepted (even by non-Catholics). Moreover, a very recent and highly intellectual pope, Benedict XVI, seems to have displayed an ambivalent attitude toward this issue; his ambivalence is emblematic and revealing, but continues to polarize and confuse. And the left-leaning social critiques tend to be summarily dismissed by practicing scientists, whose professional identity is thereby threatened, or dogmatically advocated by self-styled progressives, who seem not to have learned much from Galileo and to want to turn the clock back to pre-Galilean days.

Feyerabend portrays Galileo's trial as involving a conflict between two philosophical attitudes toward, and historical traditions about, the role of experts. Galileo allegedly advocated the uncritical acceptance by society of the views of experts, whereas the Church advocated the evaluation by society of the views of experts in the light of human and social values. Feyerabend extracts the latter principle from Cardinal Bellarmine's letter to Foscarini, asserting that "the Church would do well to revive the balance and graceful wisdom of Bellarmine, just as scientists constantly gain strength from the opinions of... their own pushy patron saint Galileo."[21] More generally, Feyerabend claims that "the Church at the time of Galileo not only kept closer to reason as defined then and, in part, even now; it also considered the ethical and social consequences of Galileo's views. Its indictment of Galileo was rational and only opportunism and a lack of perspective can demand a revision."[22]

I believe Feyerabend's thesis is untenable. In part, it is not really supported by the texts to which he refers. However, the principal difficulty is that he seems to perpetrate a fallacy of equivocation. For the principle in question could mean either that social and political leaders should evaluate the *use* of experts' views in light of human and social values, or that scientists should evaluate the *truth* of each other's views in light of human and social values.

Now, under the first interpretation, Galileo did not reject the principle, but rather would have agreed with it. Moreover, when Feyerabend attributes this principle to Bellarmine, the documentation is unclear and unconvincing. In any case, in this regard, their difference was not one of principle but of application. For example, they would have disagreed on who the relevant experts were, in particular whether theologians should be counted as experts in physics and astronomy; another disagreement would have been whether the views of theological experts should be subject to the same requirement.

Under the second interpretation, the principle was indeed rejected and criticized by Galileo. However, it is in fact untenable. For this version of the principle cannot survive the objections (which we have inherited from Galileo) against teleological and anthropomorphic ways of thinking; such thinking reduces to arguing that something is true because it is useful, beneficial, or good, and false because it is useless, harmful, or bad.

But whether untenable or not, Feyerabend's criticism is important because of the effects it has had. In fact, it has become involved in one of the latest twists to the controversy.

On the one hand, Feyerabend's apologia was politely rejected in 1989–90 by Cardinal Joseph Ratzinger, who at the time was the chairman of the Congregation for the Doctrine of the Faith (the new name of the Inquisition), and who in 2005 became Pope Benedict XVI. In a scholarly essay, in the context of an analysis of the role of faith in the revolutionary geopolitical changes happening in 1989–90, Cardinal Ratzinger quoted several anti-Galilean critiques, including Feyerabend's. However, Ratzinger went on to criticize such views as expressions of skepticism and philosophical insecurity, which he regards as unjustified.

On the other hand, there seems to be a very widespread tendency that confuses or conflates Feyerabend's view with Ratzinger's. Some authors have claimed simply that Cardinal Ratzinger or Pope Benedict *accepts* Feyerabend's view. Others have gone so far as to attribute this claim directly to Cardinal

Ratzinger or Pope Benedict XVI, without giving any indication that he was quoting Feyerabend. There have been some attempts to clarify the situation, but apparently to no avail.

In January 2008 such confusion triggered a cultural clash. A few months earlier, Pope Benedict XVI had accepted an invitation by the rector of the University of Rome to deliver the keynote address at the formal ceremony inaugurating the new academic year. This plan, however, triggered protests by students and faculty, especially in the university's distinguished department of physics. They objected primarily on the grounds of the principle of separation of Church and State, but also in part because, as they stated, they felt offended and humiliated by the pope's view of Galileo's condemnation, expressed some twenty years earlier when the pope was still a cardinal; that is, by his sharing Feyerabend's view. In the light of such opposition, and the potential for unrest and violence, the pope cancelled his speech.

This controversy is not helped, but rather exacerbated, by what seems to be a recurrent pattern of thinking or lecturing on the part of Benedict XVI, by flirting with equivocation by means of quoting controversial views. For example, an analogous issue arose as a result of a lecture he delivered at the University of Regensburg on September 12, 2006, in which he quoted a remark made by Byzantine emperor Manuel II Paleologus in 1391 regarding Islam and holy war.[23] Now, given the current geopolitical situation, Benedict did make a sustained effort to clear up the latter misunderstanding. But he made no such effort regarding the approval of Galileo's condemnation.

On the other hand, this appearance may not correspond to reality. In fact, a few months later a story surfaced in the global news media that there was a plan to erect a statue to Galileo within the Vatican walls. Without knowing more, some speculated that such a statue was a gesture to suggest that Cardinal Ratzinger had really meant to criticize Feyerabend, and that today's Church does not really approve the 1633 condemnation of Galileo. However, some time afterwards, it emerged that the Vatican statue to Galileo had been proposed by a private firm, who wanted to pay for the cost but to remain anonymous. But the latest development in this episode is that the private donor has withdrawn the offer, perhaps afraid of the unwanted publicity and controversy which the idea was generating.

However, the story did not end there. On December 21, 2008, Pope Benedict delivered the weekly Sunday speech at noon from a window of the Vatican palace to the people assembled in St. Peter's Square. Besides the usual pieties, the pope exploited the time and place to mention Galileo and the International Year of Astronomy. The pretext was provided by the fact that the feast of Christmas was originally scheduled to come around the winter solstice, and by the fact that the obelisk at the center of St. Peter's Square casts its longest shadow on the winter solstice. Then the pope went on "to greet all those who will be taking part in various capacities in the initiatives for the World Year of Astronomy, 2009, established on the fourth centenary of Galileo Galilei's first observations by telescope."[24]

As usual, such a brief reference was widely reported, commented on, and amplified. The Associated Press compiled and circulated an article entitled "Good Heavens: Vatican Rehabilitating Galileo."[25] And Church critics sprinkled the blogosphere with their share of invective and abuse.

Then on January 30, 2009, the Pontifical Council for Culture announced several projects related to the International Year of Astronomy. One project was an exhibition at the Vatican Museums entitled "Astrum 2009: The Historical Legacy of Italian Astronomy from Galileo to Today," organized jointly with the Italian Institute of Astrophysics and the Vatican Observatory, and running from October 2009 to January 2010. Another project was a conference entitled "1609–2009: From the Birth of Astrophysics to Evolutionary Cosmology," to be held in November 2009 at the Pontifical Lateran University in Rome. It is obvious that the Church was attempting to exploit the International Year of Astronomy to project an image of herself as more friendly to and harmonious with Galileo and science.

Next, on May 26–30, 2009, there was a conference in Florence entitled "The Galileo Affair: A Historical, Philosophical and Theological Re-examination." It was conceived and organized by the Florentine Jesuits, who are based at the Niels Stensen Institute in the Tuscan capital. However, its institutional sponsors represented a who's who of Italian and Vatican academic, cultural, and political institutions, including the Lincean Academy, National Research Council, Arcetri Astrophysical Observatory, Pontifical Academy of Sciences, Pontifical Council for Culture, and Vatican Observatory, as well as the President, Prime Minister, and Culture Ministry of the Italian Republic. Moreover, besides an organizing committee consisting of members and staff of the Stensen Institute, there was a scientific committee consisting of several well-known Galileo scholars. And many public and private institutions provided funding and financial support.

This conference had a very ambitious agenda, as its title suggests. The various announcements available at the website of the Stensen Institute indicated that the underlying motivation of the organizers was not only the celebration of the International Year for Astronomy, but also the continuing and lingering dissatisfaction with some aspects of Pope John Paul II's attempt at rehabilitation in 1979–92. The program included keynote addresses, presentations, and panel discussions by many distinguished scientists, historians, theologians, and philosophers. The variety of sponsoring organizations indicated that there was a general desire for fruitful dialogue not only across the divide of science and religion, but also across the separation of Church and State. The last point is especially significant, given that in Italy the Galileo affair has an additional significant complication which is absent or minor in other national contexts: the historical enmity between the Church and the political ideal of a unified Italian state.

The conference proceedings were published soon thereafter, in a volume edited by three distinguished lay scholars.[26] Unfortunately but unsurprisingly, the conference's stated ambition, initial promise, and great potential were not realized. Indeed, such events have a way of frustrating the aims and expectations of even the most astute planners and efficient organizers. In any case, dialogue between science and religion, between Church and State, and even between scholarly disciplines (such as history and philosophy) is easier said than done. Too often, instead of a real dialogue, the result is primarily a proliferation of monologues.

The year 2009 also saw the publication by the Vatican Secret Archives of a volume containing all Vatican documents relating to the trial of Galileo and dating from 1611 to 1741.[27] It was edited by the director of the Archives, Mons. Sergio Pagano. All these documents were previously known and available in print; they consist of such things as trial proceedings, minutes of Inquisition meetings, and correspondence. But their collation into a single volume was valuable and convenient for scholars. Additionally, the editor contributed a monograph-length introduction and detailed annotations. This publication contains no explicit or claimed connection with the International Year of Astronomy, and so its occurrence during that year may be just a fortuitous coincidence. On the other hand, Mons. Pagano is explicit that this volume is an expanded and improved edition of a similar but less inclusive collection which he edited in 1984; and that earlier edition was explicitly presented as a contribution to Pope John Paul's efforts at a re-examination of Galileo's trial.

After the resignation of Pope Benedict XVI, a new pope, Francis, was elected in March 2013. Pope Francis is widely known to be more pastorally inclined and less intellectually minded than his predecessor and to focus on alleviating concrete human suffering and dealing with down-to-earth problems. Thus, it is not surprising that there have been no ecclesiastic words or deeds explicitly involving the Galileo affair. However, its issues are so fundamental that they have a way of emerging in all sorts of contexts. So, for example, in 2015, Pope Francis published an encyclical entitled *Laudato Si'*, dealing with climate change. Whatever its merits, the encyclical could be (and has been) criticized for having failed to learn, from the Galileo affair, the lesson that the Church should be wary of interfering in scientific matters.

Finally,[28] the year 2016 was the four-hundredth anniversary of Galileo's first confrontation with the Inquisition. Recall that, although he was neither condemned nor tried in 1616, Cardinal-Inquisitor Bellarmine gave him a friendly warning, Commissary Seghizzi was alleged to have issued him a special injunction, and the Index condemned the Copernican doctrine as scientifically false and theologically contrary to Scripture. The Church actions of 1616 were thus instrumental in causing Galileo's trial and condemnation of 1633, and hence the subsequent and ongoing controversy.

As was to be expected, this anniversary was observed by several scholarly and cultural institutions; and so there was at least one international scholarly conference, at least one serious notice in a serious Internet publication, and at least one book exhibit in a library.[29] Of course, such observances were more in the nature of commemorations rather than celebrations. But as far as I know, there was nothing but silence on the part of the Church.

And so here we are well into the twenty-first century, still trying to come to terms with the trial of Galileo and the Galileo affair: how and why the 1633 condemnation happened; whether and why it was right or wrong; and what can be learned from it. Must this controversy continue forever? Is there not a way of resolving it?

CHAPTER 9

RELIGION VS. SCIENCE?

Conflict and Harmony in the Original Affair

The condemnation of Galileo by the Inquisition in 1633 has traditionally been interpreted as epitomizing the conflict between science and religion. This interpretation has been advanced not only by writers such as John Draper and Andrew White, whose theses have recently been widely discredited,[1] but also by such cultural icons as Voltaire, Bertrand Russell, Albert Einstein, and Karl Popper.[2]

At the opposite extreme, there is the revisionist thesis that the trial really shows the *harmony* between science and religion. This revisionist interpretation does not merely deny the traditional thesis, but reverses it. To deny the conflict thesis would mean to assert that Galileo's trial *does not* prove or illustrate the incompatibility between science and religion, or that the evidence from that trial cannot be used to justify the conflict; one would thus leave open the question of exactly what that trial indicates about the relationship between science and religion, or indeed whether the best lesson from that episode should be formulated in such terms at all. On the other hand, to reverse the conflict thesis involves retaining the terms of the question, science and religion, and arguing that the evidence from Galileo's trial, when properly examined and evaluated, supports the *opposite* of what it is traditionally taken to imply: that it supports the idea that there is a harmonious relationship between science and religion.

The most significant advocate of the harmony thesis is Pope Saint John Paul II. In his 1979 speech at the Einstein centennial, John Paul expressed his regret for Galileo's suffering "at the hands of men and organisms of the Church,"[3] and he quoted the Second Vatican Council's general condemnation of such interferences with freedom of speech and thought. He went on to give his full support for new and deeper studies of the affair, conducted in a spirit which he described as "loyal recognition of wrongs from whatever side they come."[4] Then the pope stressed that Galileo himself believed that religion and science cannot contradict each other; that the reason Galileo gave for this belief was essentially identical to that given by the Second Vatican Council; that he conducted his scientific research in the same spirit of piety and divine worship which the same council recommended as exemplary; and that he formulated important epistemological norms about the relationship between science and the Bible, which the Church later recognized as correct. John Paul summarized: "in this affair the agreements between religion and science are more numerous and above all more important than the incomprehensions which led to the bitter and painful conflict that continued in the course of the following centuries."[5]

There is no need here to reiterate the controversial, disappointing, and incomplete nature of this rehabilitation. Instead, I now want to stress that John Paul's rehabilitation of Galileo involved an attempt to elaborate the harmony between science and religion on the basis of Galileo. How successful was this attempt, and how valid is the harmony thesis?

Let's begin with the fundamental reason why the traditional conflictual view is so widely prevalent. Galileo's trial is really a series of events that began in 1613 when he wrote a letter to his former student Castelli, refuting the biblical objection to the Earth's motion—the argument that the Earth must be standing still because it is so stated or implied in many scriptural passages. One such passage is Joshua 10:12–13, where in answer to Joshua's prayer to prolong daylight during a battle, God does the miracle and stops the Sun. Galileo's letter led to an investigation by the Inquisition, which had two results in 1616. First, he was admonished to stop defending the truth of the Earth's motion. Second, the Index issued a decree which declared the Earth's motion scientifically false and theologically contrary to Scripture; temporarily banned Copernicus's book on the Earth's motion; and condemned and permanently banned Foscarini's book on the compatibility of the Earth's motion with Scripture. After Galileo was encouraged by the election of Pope Urban VIII to write and publish the *Dialogue on the Two Chief World Systems,* he was put on trial by the Inquisition and eventually found guilty of a religious crime called "vehement suspicion of heresy." And the Inquisition's sentence specified that Galileo's suspected heresy was two-fold: to hold that the Earth moves, and that it is proper to defend a scientific theory contrary to Scripture.

In light of these events, it is easy to sympathize with the traditional interpretation which summarizes the whole story by saying that here we obviously have a clash between the father of modern science and the Catholic religion. If this is not a conflict, where and when could there be one? How could one say otherwise? In particular, how can Pope John Paul II speak of harmony? I think what the pope had in mind is the following.

The harmony interpretation begins by making a distinction between the Catholic religion as such on the one hand, and men and institutions of the Church on the other. It then goes on to say that the injustices, errors, and abuses were committed by men and institutions for which they and not the Church are responsible, and so the conflict was between a scientist and some churchmen. In regard to the relationship between science and religion, the correct view is the one elaborated by Galileo himself, which the Church later adopted as its own (for example, with Leo XIII's encyclical *Providentissimus Deus* in 1893).

As we saw, that view says that God revealed himself to humanity in two ways, through His Word and through His Work. His Word, Holy Scripture, aims to give us information which we cannot discover by studying and examining His Work. But to find out what His Work is like, we need to observe it by using those parts of it which are our bodily senses and by reasoning about it with that other aspect of the Divine Work which is our mind. In short, Scripture is only an authority on questions of faith and morals, not on scientific factual questions about physical reality. In Galileo's trial, a key difficulty was the misunderstanding of these principles by the churchmen in power; once these principles are clarified and understood, as Galileo himself ironically contributed to doing, the conflict between science and religion evaporates and can subsist only in the mind or imagination of people who do not know better.

However, these considerations, although helpful, cannot be the end of the story. The situation is more complicated, and we must delve deeper.

Galileo's trial involved two main issues: the scientific issue of the location and behavior of the terrestrial globe in physical reality; and the philosophical, methodological, and theological question of the relationship between astronomical science and Scripture. The second issue reflected a disagreement between two points of view. On one side, Church officials held that Scripture is a scientific authority, and since the Copernican theory of the Earth's motion contradicts many scriptural passages literally interpreted, Copernicanism is contrary to Scripture; indeed, the 1616 Index's decree explicitly stated such a contrariety, and the 1633 Inquisition's sentence explicitly blamed Galileo for ignoring it. On the other side, Galileo held that Scripture is not a scientific authority, which implies that scriptural passages contradicting the Earth's motion should not be interpreted literally, and so Copernicanism is not really contrary to Scripture; indeed, in the letters to Castelli and to Christina, Galileo explicitly argued for such compatibility, and in the *Dialogue* he implicitly assumed it. Thus, if in this controversy we take Copernicanism to represent science and Scripture to represent religion, then Galileo was the one claiming that there is no real incompatibility between the two, whereas the Church was the one claiming that the apparent conflict between Copernicanism and Scripture is real. It follows that there was an irreducible conflictual element in Galileo's trial, between those like Galileo who believed that there is no conflict between religion and science, and those like Church officials who believed that there is a conflict. And the irony of the situation is that it was the loser or victim who held the presumably more fundamentally correct view. However, insofar as that non-conflictual view is the more plausible or nearly correct one, then it suggests a minor and residual harmonious element in the trial.

Furthermore, in more general cultural terms, Galileo was not the only one who held that there was no conflict. And many of those who agreed with him on this question of principle were themselves churchmen. For example, as we have seen, the author explicitly condemned in the anti-

Copernican decree of the Index in 1616 was Carmelite father Paolo Antonio Foscarini, who had published a book arguing that the Earth's motion is compatible with Scripture. And the author of the first published defense of Galileo was Dominican friar Tommaso Campanella.

In other words, at the time of Galileo, there was a division within the Catholic Church between those who did and those who did not accept the scientific authority of Scripture. A similar split existed in scientific circles. A further division existed in regard to the other main issue of Galileo's trial, the scientific proposition of the Earth's motion. So rather than the clash of ecclesiastic and scientific monoliths,[6] the real conflict was between two attitudes that crisscrossed both.

What should we call these two sides? How should we conceive of them? I believe the most fruitful way of describing them is to label them conservatives or traditionalists on one side and progressives or innovators on the other. The deep-structural conflict was between these two groups. In this sense, Galileo's trial illustrates the conflict between conservation and innovation and involves an episode in which the conservatives happened to win one particular battle. This conflict is one that operates across many other domains of human culture, such as politics, art, technology, and the economy. It cannot be eliminated on pain of stopping cultural development; it is a moving force of human history.[7]

However, this is not to say that the outcome is predictable, pre-determined, or inevitable. And detecting the conservative and innovative elements is not trivial. Historical agents don't come with these labels attached; and even if they did, scholars would need to judge whether such descriptions are accurate. In any case, innovators often defend their novelties by arguing that they are rooted in tradition; Galileo did this in his *Letter to Christina*, by basing his claims partly on the views of St. Augustine and other Church Fathers. Conversely, conservatives often oppose innovations by arguing that the alleged novelties are really old ideas discarded long ago; this was a common argument of the anti-Copernicans, who claimed that the Earth's motion was a Pythagorean idea that had been refuted by Aristotle in the fourth century BC and by Ptolemy in the second century AD. Finally, conservation and innovation are relative concepts whose application depends on the selection of a relevant historical period; thus, Galileo's opponents argued in part that his ideas contradicted the newest Church policies stemming from the Council of Trent (1545–63), and so represented a conservative throwback to medieval traditions.

These problems do not imply that the dialectic between conservation and innovation is a useless idea. Rather, they reinforce the need to study it concretely and contextually, without losing sight of its intended function: to help us explain the facts.

A Theological Defense of Galilean Science

A good example of the dialectic between conservation and innovation is provided by Tommaso Campanella (1568–1639). I have already mentioned him as an ecclesiastic innovator. In fact, he was a Dominican friar who had unorthodox ideas in philosophy, theology, and politics, and authored many books on these subjects. So for most of his life he was in trouble with the Inquisition; indeed, in more trouble than Galileo himself faced.

In the present context, Campanella's most relevant work is a book entitled *A Defense of Galileo*.[8] It was first published in Frankfurt in 1622, but promptly banned by the Index in Rome. Although the preface indicates that it had been written several years earlier, the exact date of composition is not given. Nor have any copies of the original manuscript survived. So specialists are somewhat divided. The account I find most plausible is the following.

Most likely, Campanella wrote his *Defense of Galileo* just before the anti-Copernican decree of March 5, 1616, and he did so at the request of Cardinal Bonifacio Caetani. Caetani was a moderate who was appointed cardinal in 1606 and member of the Congregation of the Index at the beginning of 1616, attending his first meeting on March 1, 1616. However, Caetani's request was an unofficial one. He had several Neapolitan connections that probably served as links between him and Campanella, who was in Naples serving time in prison for his unorthodox religious ideas and for his political activities advocating a kind of communist state.

The full title of the book may be translated as follows: *A Defense of Galileo, the Mathematician from Florence, where One Discusses whether the Manner of Philosophizing Advocated by Galileo Conforms or Conflicts with Sacred Scripture.* The crucial phrase here is "manner of philosophizing." Campanella is talking about Galileo's philosophical approach or manner of reasoning.[9] This contrasts with the usual translations that render the corresponding Latin phrase as philosophical view, philosophical doctrine, scientific theory, or (just) theory.

This point is extremely important because Galileo's theory, doctrine, or view (whether philosophical or scientific) suggests Copernicanism or the Earth's motion. This would imply that Campanella was more committed to the Copernican doctrine than he really was. And it would further imply that he was trying to do the same thing Foscarini had done. On the other hand, "manner of philosophizing" suggests some principle of reasoning or procedure, and so Campanella is trying to do something more general or methodological.

Campanella's broader aim is also evident from other passages and documents. For example, in the *Defense*, in the course of a main argument, Campanella states his conclusion by saying that "therefore I think that this manner of philosophizing should not be condemned,"[10] using a Latin phrase that leaves no doubt. Moreover, in a letter to Galileo, dated November 3, 1616, Campanella states that he has sent to Rome and to him a manuscript copy of his *Defense*, which he describes in Italian as "a discussion where it is proved theologically that the manner of philosophizing you use is more in conformity with Divine Scripture than the contrary one is."[11] This letter also gives a clue that Campanella's own argument is specifically or primarily a theological one.

But what Galilean manner of philosophizing is Campanella referring to? I believe this is an aspect of the manner of reasoning which Galileo uses in his discussions of astronomical topics, such as one finds in *The Sidereal Messenger* (1610) and the *History and Demonstrations Concerning Sunspots* (1613), and which he reflects on and tries to justify in his own critique of the scriptural objection, such as we find in the letters to Castelli and to Christina.

The most pertinent and general description of this manner of reasoning is to say that he advocates disregarding scriptural assertions in astronomical investigation. Stated as a methodological principle, it is the claim that scriptural statements about the Earth's rest and Sun's motion do not entail that the Earth stands still and the Sun moves. In other words, Scripture is not an authority in natural philosophy; it is the principle of limited scriptural authority. Campanella comes close to explicitly giving such a description of the Galilean approach when he says that "Galileo does not treat any of these subjects from a theological point of view, but rather by means of his marvelous instruments he renders previously hidden stars visible."[12]

In short, Campanella wants to give a scriptural argument that Scripture ought to be disregarded in scientific investigation! And besides basing this conclusion on what Scripture says, he also bases it on general theological considerations, arguing that in astronomy it is quite proper to pay no attention to scriptural assertions. In fact, such an argument constitutes a major line of reasoning in Campanella's *Defense*.

In a central part of his book, which he calls "the third assertion of the second hypothesis," Campanella stresses the fact that "in the Gospel Christ is never found to discuss physics and astronomy but only morality and the promise of eternal life."[13] Correspondingly, Campanella emphasizes two crucial scriptural passages:[14] Ecclesiastes 3:11, "God handed the world over to the disputes of men"; and Romans 1:20, "The invisible things of God come to be understood through the things which he has made." And he elaborates the point with the argument: "For us to be able to do this, he gave us a rational mind, and for avenues of investigation he provided the five senses as windows to the mind... Therefore it would have been superfluous for him, who came to redeem us from sin, to teach us what we are able and obliged to learn on our own."[15] Here Campanella is giving a justification of the principle of limited scriptural authority as being implicit in Jesus' example in Scripture, explicit in the assertions of the Old and the New Testaments, and in accordance with plausible theological speculation.

But Campanella goes further. He argues not only that it is proper to learn about the world by using our mind and senses rather than by reading

Scripture, but also that it is un-Christian to prevent such learning. In what he calls "the fourth assertion of the second hypothesis," Campanella holds that "anyone who forbids Christians to study philosophy and the sciences also forbids them to be Christians."[16] One reason is that "since one truth does not contradict another, as was stated by the Lateran Council under Leo X and elsewhere, and since the book of wisdom by God the creator does not contradict the book of wisdom by God the revealer, anyone who fears contradiction by the facts of nature is full of bad faith."[17] Another is that

> from the beginning the world has been called the "Wisdom of God" (as was revealed to St. Brigid) and a "Book" in which we can read about all things. Hence, in his Sermon 7 on the fast days of the tenth month, St. Leo says, "We understand the meaning of God's will from these very elements of the world, as from the pages of an open book".[18]

It follows that "therefore wisdom is to be read in the immense book of God, which is the world, and there is always more to be discovered."[19] Using the metaphor of the book of nature, and the theological claim that this book was authored by God and so is at least as important as the book of Scripture, Campanella is arguing that it is wrong (theologically) to prevent someone from reading the book of nature.

Similarly, in what he calls "the first assertion of the second hypothesis," Campanella argues that it is not only irrational and harmful but impious "if there is anyone who chooses on his own to prescribe rules and limits for philosophers as though they were decreed in the Scriptures and who teaches that one should not think differently than he does, and who subjects and confines the Scriptures to one unique meaning either of his own or of some other philosopher."[20] This impiety looks curiously like a description of what Galileo's scriptural critics were doing, and the reason Campanella gives is something which he also finds in both St. Augustine and St. Thomas Aquinas. Campanella's reason is that such an impious person "exposes the Sacred Scriptures to the mockery of the philosophers and to the ridicule of pagans and heretics and thereby prevents them from listening to the faith."[21] And then he gives a long quotation from St. Thomas that includes a quotation from St. Augustine's *On the Literal Interpretation of Genesis*. Thus, Campanella is able to claim that "so says St. Thomas in agreement with St. Augustine,"[22] and hence to provide formidable theological credentials to his own argument.

When it comes to the Joshua passage, Campanella denies that the miracle "would be nullified if the sun is at rest in the center of the world."[23] For "the appearances are exactly the same if either the observer or the object seen is moved,"[24] and the Copernicans say that the miracle happened by stopping the Earth rather than the Sun; but "whoever says that this happened by arresting the motion of the earth does not deny the miracle but explains it, just as the physicist does not deny that God causes the rainbow but explains how he does it and what natural and reasonable means he uses."[25] Campanella does not seem worried about the nonliteral interpretation that is needed, but about retaining the spiritual message or meaning.

Campanella ends his book with the following profound and prophetic formulation of his conclusion:

> In my judgment, in agreement with what St. Thomas and St. Augustine have taught us in our Second Hypothesis, it is not possible to prohibit Galileo's investigations and to suppress his writings without causing either damaging mockery of the Scriptures, or a strong suspicion that we reject the Scriptures along with heretics, or the impression that we detest great minds... It is also my judgment that such a condemnation would cause our enemies to embrace and honor this view more avidly.[26]

Compare Campanella's position with those of Galileo and Foscarini. In his *Letter on the Earth's Motion*, Foscarini criticized the scriptural objection by saying that Copernicanism is not contrary to Scripture because, although contrary to its literal meaning, Copernicanism is a thesis about natural phenomena; but Scripture does not aim to make claims about nature; and so scriptural statements about natural phenomena are to be interpreted not literally, but rather in accordance with the principle of accommodation.

In his *Letter to Christina*, Galileo argued that literal meaning, patristic interpretation, and scriptural authority are all irrelevant for demonstrable questions of natural philosophy, although they are relevant or binding for matters of faith, morals, history, and indemonstrable claims about nature. Such irrelevance follows from the universally accepted principle of the priority of demonstration, which in turn follows from the two-fold aspect of divine revelation and the asymmetries between the Work and the Word of God. Thus, both steps of the scriptural argument are non-sequiturs: the inference that the Earth's motion is contrary to Scripture because it is contrary to its literal meaning and patristic interpretation; and the inference that because it is contrary to Scripture, therefore it is false.

Moreover, the literal meaning of Joshua really contradicts the Ptolemaic system; whereas Copernicanism (in the modified Galilean version that attributed axial rotation to the Sun) is largely compatible with the literal interpretation of Joshua. So the key premise of the biblical argument is questionable or false.

In his *Defense of Galileo*, Campanella argued that the assumption that Scripture is a scientific authority is itself contrary to Scripture, as well as deeply un-Christian, and contradictory to the patristic tradition of St. Augustine and St. Thomas.

So, Foscarini advanced a theological criticism of the minor premise of the scriptural objection—that Copernicanism is contrary to Scripture. Campanella put forth a theological criticism of its major premise—that Scripture is a scientific authority. Galileo proposed a methodological criticism of this major premise, and a scientific and textual criticism of the minor premise. Together, Foscarini, Campanella, and Galileo provide us with theological *and* philosophical arguments justifying the claim that Copernicanism is not contrary to Scripture, and that Scripture is not a scientific authority.

In my judgment, these arguments of Foscarini, Campanella, and Galileo are cogent and essentially valid. They are thus of perennial relevance and have some applicability to subsequent and present-day issues regarding the relationship between science and religion.

Two final remarks. If my judgment is right, then the original Galileo affair is an episode during which a significant development in natural philosophy or science (the Galilean telescopic discoveries and the re-assessment of

Copernicanism) produced a significant development in biblical hermeneutics and theology (the discovery of reasons why Scripture is not a scientific authority); or at any rate the beginning of the modern establishment of this approach to Scripture.

Yet despite the cogency of these criticisms and arguments, we know that they failed to convince the officials of Catholic Church. So, secondly, this case represents one of the greatest ironies in the history of the interaction between science and religion. At the intellectual level, we have the invention and discussion of some of the best arguments ever advanced why a particular scientific theory was compatible with Scripture and why in general Scripture is not a scientific authority. But at the institutional level, one of the world's great religions issued a formal condemnation of a key scientific theory that played a crucial role in the rise of modern science. A high point in the history of thought was accompanied by a low point in the history of action.

Conflict and Myth in the Subsequent Affair

The story still does not end here, not yet. For, in the case of the Galileo Affair, there are complications stemming from what happened subsequently.

Even those who advocate the harmonious account of the original trial do not deny that the key feature of the subsequent Galileo affair has been indeed a conflict between science and religion. In fact, as we saw earlier, Pope Saint John Paul II believed that the lesson from Galileo's original trial is the harmony between science and religion, and he wanted to stress and elaborate this lesson in order to try to put an end to the subsequent, very real, but presumably unjustified science-versus-religion conflict. The science-versus-religion conflict is indeed an essential feature of the subsequent controversy, even more integral to it than to the original trial. But underneath lies a deeper conflict—that between cultural myths and documented facts. Let me elaborate.

It may be true, as Pope John Paul and many other authors have advocated, that the animosities between science and religion are a thing of the past, and that there is no longer any good reason for them to continue. According to their argument, during the Enlightenment the view was developed that the trial of Galileo embodied the inherent incompatibility between science and religion, and later this view became widely accepted. They also argue that this view was the result of inadequate historical knowledge about the trial and of philosophical and ideological biases. For example, it overlooks the pro- and anti-Galilean split within Catholicism at the time of the original affair, and the crucial fact that despite the opposition Galileo experienced at the practical level, at the reflective level he himself believed in the harmony between science and religion, and (in the letters to Castelli and Christina) gave very good arguments to justify such harmony. And the view also presupposes the Platonist idea that science and religion are eternal unchanging self-subsisting entities which by definition have a certain nature that places them at war with each other, rather than being historical dynamic entities that are sometimes at war and sometimes in harmony.

However, although this pro-harmony argument is important for a full understanding of the *original* affair, it does not undermine the essential correctness of the *subsequent* conflict.

Historically speaking, the view of Galileo's trial as epitomizing the conflict between science and religion was not an Enlightenment invention (let alone the invention of nineteenth-century authors Draper and White), but started to be developed immediately after the 1633 condemnation. This happened when (in 1633–6) an international group of liberal-minded secularists translated into Latin and published in Strasbourg[27] Galileo's banned *Dialogue* and the incriminating *Letter to Christina*.

Also, even if the conflictual view of Galileo's trial were incorrect, and a thing of the past that should now be replaced by the harmony view, it would be naïve and wrong to deny the truth and consequences of the fact that for about four centuries such an incorrect view has been the most popular interpretation of the episode.

In any case, the Platonist, static conception of science and religion may be inadequate, and so perhaps science and religion are not necessarily and always in conflict, and there may be many episodes when they have been in harmony. However, the case of Galileo may have been one where science and religion happened to be in conflict.

In fact, as we saw earlier, Galileo's trial *does* exhibit such a conflict if in that context we equate science with Copernicanism and religion with Scripture; for although Galileo himself believed and argued that Copernicanism was compatible with Scripture, his opponents (Bellarmine, Urban, the Index, and the Inquisition) all claimed that Copernicanism was contrary to it. Hence, there is an irreducible historical conflict there.

Next, it is important to stress that the conflict between science and religion is a striking feature of both the original and the subsequent Galileo affair: in the original episode it takes the form of Copernicanism versus Scripture; in the subsequent controversy it appears as the perception that the trial of Galileo epitomizes the conflict between science and religion. The important difference involves the underlying structure. In the original affair, that deep structure is the conflict between conservation and innovation, which generated the split within the Catholic Church (and within Protestantism, within astronomy, and within natural philosophy) about the relationship between Copernican astronomy and scriptural interpretation. In the subsequent controversy, the deep structure lies in the conflict between myths and facts—the rise, evolution, and fall of cultural myths; the trial of Galileo became a great occasion for mythologizing not only on the part of anti-clerical and pro-Galilean elements, but also (somewhat reactively) on the part of pro-clerical and anti-Galilean forces.

In fact, myth-makers on both sides have been busy for four centuries. For example, on the anti-clerical side, in the *Areopagitica* (1644) John Milton recalled his visit to Tuscany in 1638–9 and his meeting with Galileo, remarking, "there it was that I found and visited the famous Galileo grown old, a prisoner to the Inquisition, for thinking in astronomy otherwise than the Franciscan and Dominican licensers thought."[28] In the *Essay on the Customs and Spirit of Nations* (1753), Voltaire commented,

> In a decree issued in 1616, a congregation of theologians declared Copernicus's opinion, so well brought to light by the Florentine philosopher, "not only heretical in the faith, but also absurd in philosophy". This judgment against a

> truth later proved in so many ways is clear testimony of the force of prejudice. It should teach those who have nothing but power to be silent when philosophy speaks and not to interfere by deciding what is not within their jurisdiction. Then in 1633, Galileo was condemned by the same tribunal to prison and to do penance, and he was obliged to recant on his knees. In truth, his sentence was milder than that of Socrates; but it was no less disgraceful to the reason of the judges of Rome than the condemnation of Socrates was to the enlightenment of the judges of Athens.[29]

Nearly a century later, in 1841, in a book widely translated and circulated in Italian, French, and German, the Italian polymath Guglielmo Libri (1803–69) concluded his account with these words:

> The persecution of Galileo was odious and cruel, more odious and more cruel than if the victim had been made to perish during torture. For . . . the Inquisition . . . was not merely after Galileo's body; they wanted to strike him morally; they forbade him to make discoveries. Enclosed in a circle of iron, blind, and isolated, he was left to be consumed by the anguish of a man who knows his strength but who is prevented from using it. This ill-fated vengeance, which Galileo had to endure for such a long time, had the aim of silencing him; it frightened his successors and retarded the progress of philosophy; it deprived humanity of the new truths which his sublime mind might have discovered. To restrain genius; to frighten thinkers; to hinder the progress of philosophy; that is what Galileo's persecutors tried to do. It is a stain which they will never wash away.[30]

And in 1953 Albert Einstein, writing the Foreword to an English translation of the book (*Dialogue*) which occasioned Galileo's condemnation, expressed this judgment: "A man is here revealed who possesses the passionate will, the intelligence, and the courage to stand up as the representative of rational thinking against the host of those who, relying on the ignorance of the people and the indolence of teachers in priest's and scholar's garb, maintain and defend their position of authority."[31]

On the anti-Galilean side, for a while after the condemnation, there was an attempt to discredit Galileo's ideas by taking his abjuration at face value. A good example is provided by a passage in English author Alexander Ross's *The New Planet no Planet* (1646). This book was a rebuttal to John Wilkins's *A Discourse concerning a New Planet, Tending to Prove that 'tis Probable*

Our Earth Is One of the Planets (1640). Taking issue with Wilkins's claim that many astronomers followed Copernicus, Ross retorted: "And yet of these five you muster up for your defense, there was one, even the chiefest, and of longest experience, to wit, Galileus, who fell off from you; being both ashamed, and sorry that he had been so long bewitched with so ridiculous an opinion."[32]

In 1784, in France, Jacques Mallet du Pan started a myth already mentioned but worth repeating: "Galileo was persecuted not at all insofar as he was a good astronomer, but insofar as he was a bad theologian."[33] The bad theology which Mallet misattributed to Galileo was the use of Scripture to prove astronomical propositions—the opposite of what Galileo preached and practiced. Less absurd versions of this myth claim that Galileo was a bad theologian in the sense of being a non-theologian who intruded into hermeneutical controversies.

In *The Martyrs of Science, or the Lives of Galileo, Tycho Brahe, and Kepler* (1841), Scottish physicist David Brewster portrayed Galileo as a coward:

> In the ignorance and prejudices of the age—in a too literal interpretation of the language of Scripture—in a mistaken respect for the errors that had become venerable from their antiquity—and in the peculiar position which Galileo had taken among the avowed enemies of the church, we may find the elements of an apology, poor though it be, for the conduct of the Inquisition. But what excuses can we devise for the humiliating confession and abjuration of Galileo?... Galileo cowered under the fear of man, and his submission was the salvation of the church. The sword of the Inquisition descended on his prostrate neck; and though its stroke was not physical, yet it fell with a moral influence fatal to the character of its victim, and to the dignity of science.[34]

And we have already seen that Pierre Duhem, in *To Save the Appearances* (1908), tried to portray Galileo as a bad logician and epistemologist, while Paul Feyerabend, in *Against Method* (1988), considered the Church's indictment of Galileo as rational for the time.[35]

What I am claiming is that, whereas the dialectic of conservation and innovation forms the deep structure underlying the science–religion conflict of the original affair, the just described two-sided myth-making forms the deep structure underlying the science–religion conflict of the subsequent Galileo affair. Furthermore, although bilateral, it is obvious that such myth-making is not otherwise symmetric: the pro-Galilean and anti-clerical myths tend to be aggressive, while the pro-clerical and anti-Galilean ones tend to be defensive; and this is opposite to the asymmetry in the conflict in the original affair, in which "science" (through Galileo) was the apparent victim, and "religion" (through the Catholic Church) was the apparent aggressor. Finally, besides the important substantive issues raised by the various views of Galileo's trial, the development of such cultural myths deserves study in its own right. Such a study must be balanced, judicious, bipartisan, and objective with respect to the pro- and anti-clerical and pro- and anti-Galilean dichotomies; this, of course, is easier said than done. Let us illustrate such issues by means of an example.

The Villa Medici Myths

Villa Medici in Rome is one of the most impressive palaces in the city, and for a long time it was the property of the Grand Duchy of Tuscany, ruled by the House of Medici (see Figure 20). This palace was not the Tuscan embassy; the residence and office of the Tuscan ambassador to the Holy See was Palazzo Firenze, closer to the center of Rome.[36] Rather, Villa Medici was a palace where members of the Medici family and other special guests could reside while visiting Rome; the property had a large adjoining garden, so that such visits would not be too stressful.

After 1610, when Galileo had received the title of "Philosopher and Chief Mathematician to the Most Serene Grand Duke of Tuscany," he often resided at Villa Medici during his visits to Rome. His last stay at the villa was in 1633, after the sentencing at the trial, when he was kept under house arrest there for about a week.

Next to the building, at the edge of the street stands today a commemorative column, erected at the end of the nineteenth century, which reads as follows: "The palace next to this spot, / which belonged formerly to the

Figure 20. Villa Medici, Rome

Medici, / was a prison for Galileo Galilei, / guilty of having seen / the earth turn around the sun / SPQR / MDCCCLXXXVII."

The origin of this monument goes back to 1872, immediately after the city of Rome ceased to be the capital of the Papal State and became the capital of a united Italy, known as the Kingdom of Italy.[37] The municipal government of Rome wanted to affix an inscription on the facade of the palace in memory of Galileo. However, the building then belonged to the French government, which opposed the project, feeling that the monument would offend the pope. Thus, it took another 15 years before the monument came into being, and it had to take the form of a free-standing commemorative column on public property in the adjacent street.

The event was applauded by the anti-clerical press, but it was sharply criticized by the Vatican newspaper, in an unsigned article entitled "Epigraphs and Insults." It objected that Galileo was not "imprisoned" in Villa Medici; that he was not found guilty of "having seen the earth turn around the sun"; but that his transgression was to have attempted "to change the physical and astronomical question into a theological one."[38]

There is clearly a clash here between science and religion, or at least between some people's perceptions of science and some institutional practices of religion. But the mythological aspect of the episode is even more basic.

To begin with, the Villa Medici column echoes several common myths about the original Galileo affair. One involves the unqualified talk of imprisonment; this is a falsehood if meant literally, and an equivocation if meant figuratively to refer to house arrest. A second pertains to the one-sided focus on the astronomical issue, which disregards the methodological issue of whether Scripture is an astronomical authority. More generally, the inscription suggests the incompatibility between science and religion.

More strikingly, however, the Villa Medici inscription is a good formulation of the empiricist myth, through its talk of Galileo having seen the Earth turn around the Sun. Again, perhaps this is merely an equivocation, trading on the ambiguity between the literal and the figurative meanings of the notion of seeing. But it is more likely that the expression is meant literally, because of the construction of the original Italian sentence, because of how the sentence was interpreted by the Vatican newspaper in its criticism, and because the creators of the inscription probably meant to refer to telescopic observation. Interpreted literally, thc statement is erroneous, since it is impossible to observe the Earth's motion directly, even today, let alone in Galileo's time.

Such empiricism is a myth in the sense that it reflects the principle that observation is paramount in science. This principle is strictly speaking false, and yet many working scientists pay lip service to it; so believing it seems to perform a good function in science. In saying this, I am borrowing the concept of myth elaborated by professional scholars of mythology:[39] they stress that myths are beliefs which are literally not true, but which often contain some partial truths, and which, more importantly, perform valuable and important functions in defining and preserving the cultural cohesiveness of social groups. Of course, the partial truth of the empiricist myth is that the astronomical discoveries which Galileo made by means of telescopic observation were crucial and indispensable for his own re-assessment of the geokinetic theory and consequently for the Copernican Revolution.

On the other hand, in opposing this myth the religious side succumbed to its own anti-Galilean myth-making, one of these being the claim originating with Mallet du Pan that Galileo was condemned not for being a good astronomer, but for being a bad theologian. The inventors and perpetrators of this myth are misled by the fact that on various occasions Galileo engaged in biblical exegesis, arguing that the biblical texts adduced by his critics against the Earth's motion are more in accordance with the Copernican than the Ptolemaic system. But such Galilean exercises at biblical exegesis are taken out of context. The context of Galileo's biblical exegesis makes it clear that he is not advocating the principle that it is proper to support scientific claims by means of Scripture. As we have seen, in truth, Galileo held the opposite principle.

However, such a misinterpretation of Galileo's view may be taken to help many Catholics make sense or justify the Inquisition condemnation, thus enhancing their religiosity and piety. This is precisely the function of cultural myths, at least according to one theory advanced by scholars of mythology.

So far, I have identified the relevant myths in the episode of the Villa Medici monument. Equally important is the examination of the interaction between such myths and facts, namely documented facts, as established by historians and scholars. Thus, I would want to note that the two particular myths in question seem to have been eventually discarded.

Nowadays, no serious spokesman for science would claim that fundamental facts like the Earth's motion can be observed directly. Einstein and Popper both emphasized the conceptual aspects of the scientific method. Similarly, no churchman would today attribute to Galileo the view that Scripture is a scientific authority. Recall that in the recent Vatican re-examination of the Galileo affair, John Paul II was clear that Galileo not only held the correct principles about the relationship between science and Scripture, but also gave insightful reasons in support of such principles; i.e., that Galileo was a good theologian. And these papal pronouncements only make explicit what was implicit in the encyclical *Providentissimus Deus*, by Pope Leo XIII in 1893.

On the other hand, with regard to the theory versus observation issue, now the pendulum seems to have swung to the other extreme: a common slogan nowadays is that all observation is theory-laden,[40] which makes theorizing paramount in scientific research. Partly as a result, a new portrayal of Galileo has been developed, by Feyerabend and others, as someone who not only had no observations to prove his views and refute his opponents, but did not even have good arguments; instead, he was allegedly an epistemological anarchist for whom anything goes, or a sophistical rhetorician who had mastered the art of making the worse argument appear stronger. A new myth about Galileo's trial has arisen or been revived.

Interestingly, something analogous has happened on the religious side. The pro-Galilean efforts by John Paul have given rise to the widespread belief that the Church has officially rehabilitated him. As we saw earlier, this did not in fact happen. And as some perceptive scholars have written,[41] this belief is really the very latest myth in an ongoing and perhaps unending story. Moreover, in the course of the rehabilitation efforts, Pope John Paul and Cardinal Poupard succumbed to Duhem's myth that Galileo's ecclesiastic opponents were better methodologists than he was.

In the subsequent Galileo affair, then, the conflict takes the form of science vs. religion, and is pervasive, overwhelming, and undeniable. Pro-clerical spokesmen (like John Paul II) may bemoan this fact, but do not deny the historical existence and reality of this subsequent conflict. I am not denying it either, but I have tried to explain it on the basis of something I regard more fundamental. Taking both science and religion as important elements of culture, I try to identify the myths that are operative in them; as in other cultural institutions, myths play a significant role.

CHAPTER 10

A MODEL OF CRITICAL THINKING?

Learning from Galileo

From prehistoric times until the beginning of the sixteenth century, almost all scientists and philosophers believed that the Earth stands still at the center of the universe and all heavenly bodies revolve around it. By the end of the seventeenth century, most scientists and philosophers had come to believe that the Earth is the third planet circling the Sun once a year and spinning around its own axis once a day. The transition was a slow, difficult, and controversial process. We may fix its beginning with the publication in 1543 of Nicolaus Copernicus's book *On the Revolutions of the Heavenly Spheres*, and its completion with the publication in 1687 of Isaac Newton's *Mathematical Principles of Natural Philosophy*.

The discovery that the Earth moves and is not situated at the center of the universe involved not only key astronomical and cosmological facts, but was interwoven with the discovery of the most basic laws of nature and principles of physics: the law of inertia, the relation between force and acceleration, the law of action and reaction, and the principle of universal gravitation. It was also connected with the clarification and improved understanding of key principles of scientific method. It represents, therefore, without a doubt the most significant breakthrough in the whole history of science. Accordingly, the series of developments that started with Copernicus in 1543 and ended with Newton in 1687 is sometimes labeled the Scientific Revolution.

More generally, it would perhaps be no exaggeration to say that this transition represents the most important intellectual transformation in human history.[1] One reason for this involves the world-wide repercussions of the Scientific Revolution itself: science seems to be the only cultural force that has managed to penetrate and dominate human societies and cultures in all parts of the Earth. Another reason involves the broad interdisciplinary impact of the transition from a geocentric to a geokinetic world view, which had profound effects not only on the many branches of science, but also on philosophy, theology, religion, art, literature, technology, industry, and commerce; indeed, it changed mankind's self-image in general. We may thus also call this transition the Copernican Revolution, if we want a label that leaves open its broad ramifications outside science; this label also gives due credit to the one thinker whose contribution initiated the process.

As we have seen, Galileo made essential and crucial contributions to the Copernican Revolution. His trial and the original Galileo affair hinged precisely on the key scientific claims of the Copernican world view and the corresponding methodological issues. So it is only natural to try to derive from these events general lessons about scientific method, critical thinking, and human rationality. One would thereby be using Galileo as a kind of model. Indeed, scientists like Isaac Newton, Albert Einstein, and Stephen Hawking have tried to derive such methodological inspiration from the Galilean model; and the same has been done by philosophers like David Hume, Immanuel Kant, and José Ortega y Gasset.[2]

Fallibility and Reasonableness

One of the things that makes the Copernican Revolution so relevant to human rationality stems from the fact that it involves some beliefs which today are known with certainty to be false and incorrect, and others which are now established with equal conclusiveness as being absolutely true and correct. No sane person today would question the fact that the Earth moves. If human knowledge encompasses any item of information, then the Earth's motion is surely one such. Conversely, if we know anything to be false, it is the idea that the Earth stands still at the center of the universe.

These epistemological facts have two sets of implications that point in opposite directions. On the one hand, there are positive lessons to be drawn. One is that knowledge is possible because it is actual, and it is actual because we know at least one thing, namely that the Earth moves and is not standing still at the center of the universe. Another positive lesson is that progress is possible because the Copernican Revolution is an instance of it; and this is so because the result was to replace ignorance by knowledge in regard to the question of the location and the behavior of the Earth in the universe.

However, there are also lessons that might be called negative, and these are the ones that point more explicitly in the direction of human rationality. To see this, we must focus on the fact that for thousands of years, until relatively recently in human history, almost everybody was wrong about a very fundamental matter; and this included scientists and philosophers, the supposed experts. So it is possible for everyone to be wrong, or at least for everyone to be wrong for some time, and even for a long time; for that was certainly the case in regard to the motion of the Earth up until the time of the Copernican Revolution.

This lesson can give encouragement to would-be critics, no matter how radical; it is possible they may be the only persons to see the truth about the topic in question. Be that as it may, the point is that it is always possible that almost everyone is wrong about almost anything. But this is just one side of the lesson relevant to human rationality, the one that concerns the element of what may be called *criticism* or *fallibility*. There is another element, which we may call *reasoning* or *reasonableness*.

If we do not neglect the element of reasonableness, then we are led to ask the following about the Copernican Revolution. Although subsequently found to be factually wrong in the content of their belief, were pre-Copernicans reasonable or unreasonable in holding their incorrect belief? In other words, was their reasoning sound or unsound?

In fact, the reasoning of the pre-Copernicans in believing that the Earth stands still at the center of the universe was essentially correct. We have already seen the many arguments for the geostatic view and against Copernicanism. So the Copernican Revolution does not demonstrate that it is possible for everyone to be unreasonable. On the contrary, it suggests that at the level of reasoning, humankind is essentially reasonable. Another lesson of the Copernican Revolution is, then, that it is possible for almost everybody to hold a false belief, but not for almost everyone to be unreasonable.

At this point some people would begin to despair about how the transition to a geokinetic view was ever possible, and they may think that the transition was itself unreasonable or irrational. This conclusion does not follow, for at least three reasons. First, the reasonableness of the arguments against the Earth's motion does not mean that they were perfectly right or totally correct; it was quite proper to look for their weak points. Second, the situation was not static, but evolved when new discoveries were made. And third, although the criticism of opposite arguments that are reasonable is difficult, it can be done in a fair-minded way. We'll consider each of these points below.

Reasoning and the New Physics

As we have seen, Galileo was at first primarily interested in physics and mechanics, and was working on a research program designed to understand in general how bodies move. He was critical of the ancient physics of Aristotle and was developing a new theory of motion more in line with the work of another ancient Greek—Archimedes. Galileo was aware of Copernicus's new argument, but felt its insufficiency and the greater power of the many anti-Copernican and pro-geostatic arguments. He was initially attracted to the Copernican theory because its key geokinetic hypothesis was more in accordance with the new physics he was developing. In fact, his new physics provided him with an effective criticism of the mechanical objections to the Earth's motion, and with some physical evidence in its favor. The connection can be seen most clearly and most simply for the case of the vertical fall objection.

Recall that the vertical fall objection argued that the Earth cannot rotate because on a rotating Earth freely falling bodies would have no reason to keep up with the Earth's motion, and hence during free fall they would be left behind; this in turn means that they would be falling in a slanted direction and not vertically; but it is obvious that they do fall vertically. Here the last step in reasoning can be reconstructed as an instance of *denying the consequent*. This is the argument-form consisting of two premises and one conclusion, such that one premise (called the major premise) is a conditional ("if–then") proposition, the other premise (called the minor premise) denies the consequent ("then") clause, and the conclusion denies the antecedent ("if") clause:

(1) if the Earth rotated, then bodies would not fall vertically;
(2) but bodies do fall vertically;
(3) so the Earth does not rotate.

Galileo begins his critique by asking us to focus our attention on the meaning of the proposition that bodies fall vertically. What does it mean? What is meant by vertical fall? Does it mean fall from the terrestrial point where the body is released to the point on the Earth's surface directly below, such as the motion from the top to the base of a tower? Or does vertical fall mean fall along the straight line in absolute space going from the point of release to the center of the Earth? In other words, does vertical fall mean fall perpendicular to the Earth's surface as viewed by a terrestrial observer (standing on the Earth's surface), or as viewed by an extraterrestrial observer (looking at the whole globe from a fixed point at a distance)? Let us call the first *apparent* or *relative vertical fall*, and the second *actual* or *absolute vertical fall*. These are indeed different.

To understand this difference, consider Figure 21. The semicircles on the left side represent portions of the Earth's equator, and the structures labeled AB and A´B´ represent towers on the Earth's surface. The parts of the figure on the right side are simply highly magnified representations of the situations on the left, so that the Earth's surface appears flat because the distance involved (BB´) is very small. Figure 21a represents a motionless Earth, whereas Figures 21b and 21c represent the Earth undergoing axial rotation

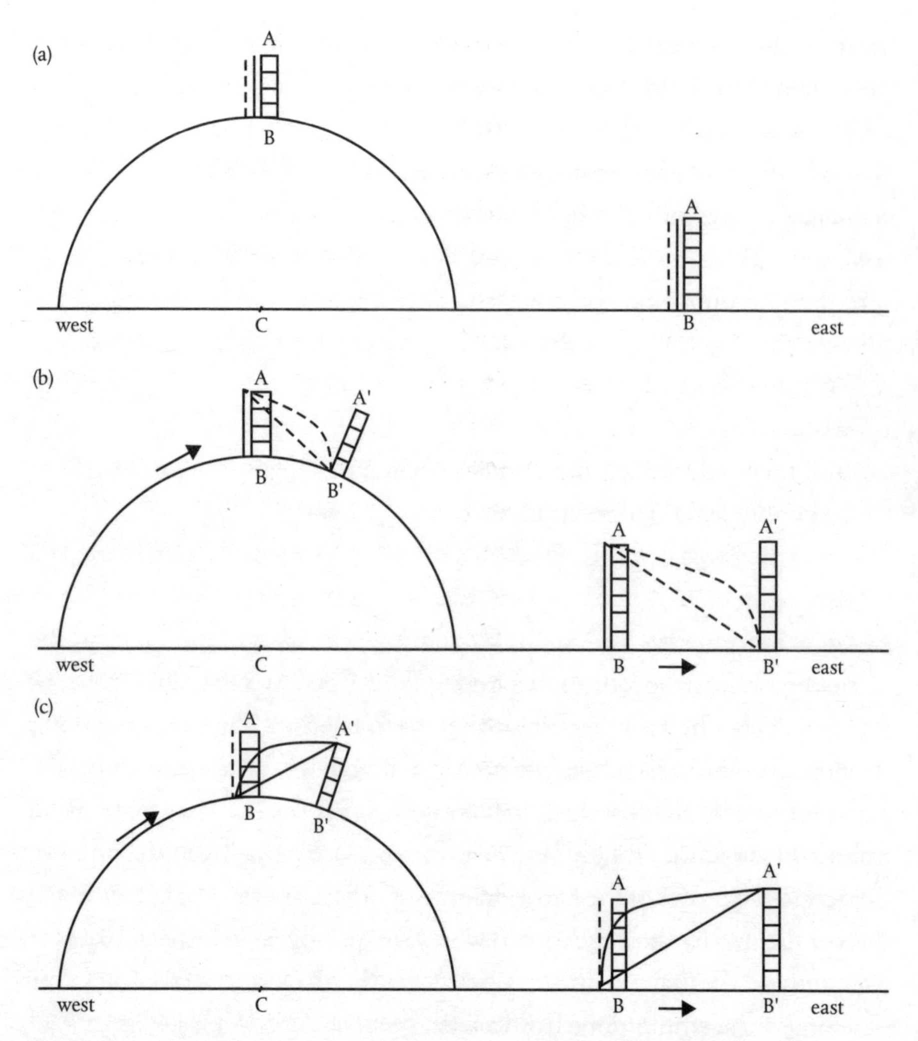

Figure 21. Apparent vs. actual vertical fall on motionless vs. rotating Earth

from west to east (or clockwise), as suggested by the arrows. Figures 21b and 21c show two different positions of the tower: the *unprimed* position (AB) represents the tower's position at the beginning of the experiment of dropping a rock from the top of a tower and letting it fall freely; the *primed* position (A´B´) represents the tower's position at the end of the experiment when the fallen rock has reached the ground. The solid lines represent apparent vertical fall, and the dotted lines actual vertical fall. The lines (whether solid or dotted) between towers are drawn both as straight slanted lines and as parabolic slanted lines; here the main point to note is that they are both slanted, and although the parabolic representation is more accurate, this refinement plays no role in this discussion. In Figure 21a representing a motionless Earth, apparent and actual vertical fall coincide. In Figure 21b, in which the Earth is rotating, apparent vertical fall is experienced on Earth, but actually slanted fall takes place. In Figure 21c, also representing a rotating Earth, actual vertical fall is taking place, but apparently slanted fall is experienced on Earth.

With these pictures in mind, we are now in a position to better follow Galileo's reasoning. To explain the difference between actual and apparent vertical fall, Galileo points out that although apparent and actual vertical fall would coincide on a motionless Earth (Figure 21a), they would *not* coincide on a rotating Earth (Figures 21b and 21c).

Assume the Earth were in rotation and a rock is dropped from the top of a tower (A). If the rock *appeared* to fall vertically to a terrestrial observer (Figure 21b), then it would be seen to land at the foot of the tower (B´). But, on a rotating Earth, the foot of the tower would have undergone some rotational motion during the time of fall, and so as viewed by an extraterrestrial observer the *actual* path (AB´) of the rock would be slanted toward the east. So on a rotating Earth, apparent vertical fall would not produce actual vertical fall, but rather actually slanted fall.

Similarly, given the same assumption of terrestrial rotation, if the rock were to move with *actual* vertical fall (AB in Figure 21c), then it would land to the west of the base of the tower because the base of the tower would have moved eastward to point B´. Therefore, to a terrestrial observer the path of the falling rock would *appear* slanted westward, because when the rock reached point B the terrestrial observer would have moved to A´, and so the *apparent* path would be A´B, which is not vertical. So, on a rotating Earth, actual vertical fall would not produce apparent vertical fall, but rather apparently slanted fall. This contrasts with a motionless Earth, for which if the path were from the top to the base of the tower as seen by the terrestrial observer, then it would also be straight and perpendicular to the Earth's surface for the extraterrestrial observer, and vice versa (Figure 21a).

Having made such a distinction, Galileo applies it to the argument above. Suppose that, when the argument claims that the Earth cannot rotate because bodies fall vertically, the vertical fall in question is actual vertical fall. Then we would be entitled to ask how you know that bodies do actually fall vertically, for observation on the Earth only reveals *apparent* vertical fall. While it is undeniable that to us bodies are seen to fall from the top to the base of a tower, we have no experience about how their paths look from an extraterrestrial viewpoint. How could one answer that question? How could one justify that falling bodies move with actual vertical fall?

It seems that one could only try an empirical justification, by basing actual vertical fall on apparent vertical fall. But to do this would presuppose that apparent vertical fall implies actual vertical fall, and in turn this implication amounts to assuming that the Earth is motionless, since this is the only condition under which the implication holds. Unfortunately, the motionlessness of the Earth is the very conclusion the argument is trying to prove. In short, interpreted in terms of actual vertical fall, the objection from vertical fall begs the question because the premise that bodies fall vertically is either assumed gratuitously or supported circularly (by reasons that presuppose the conclusion).

But perhaps the objection from vertical fall intends *apparent* vertical fall. Then, the minor premise of the above-mentioned argument would be both true and uncontroversial. In this case, Galileo questions the *major* premise, the conditional proposition that if the Earth were in rotation then bodies would not undergo *apparent* vertical fall. What are the grounds for asserting this conditional claim?

At that time, the justification of this claim was based on some basic principles of Aristotelian physics. One was the principle that a body can have only one natural motion; another was the principle that the natural state of heavy material bodies is rest; a third was that motion requires a force to sustain it. To understand the connection, we have to understand the first answer the Aristotelians would give in this discussion.

On a rotating Earth, if bodies appeared to fall vertically, then in reality they would follow a path slanting eastward (Figure 21b), the resultant of downward and horizontal components. Now, according to Aristotelian physics such a mixture or combination is impossible because the horizontal component of motion would be motion under the influence of no external force, and so it would have to be natural; but such a second natural motion could not coexist with the first, downward motion.

The issue then becomes whether it is possible for a free-falling body to have a horizontal component of motion. This is where some of the principles of the new Galilean physics come in; they are the principle of the conservation of motion and the principle of the superposition or composition of motions. Conservation of motion is an approximation to such laws of modern physics as the law of inertia, the law of conservation of linear momentum, and the law of conservation of angular momentum. The Galilean formulation relevant here is that if a body is moving horizontally, then it will conserve its motion as long as it is left undisturbed. And the principle of superposition asserts that it is physically possible for a body to have more than one tendency to move; and in these cases the actual motion will be the resultant, as defined by the diagonal of the corresponding parallelogram.

These principles can now be applied to answer the objection from *apparent* vertical fall. If the Earth rotated, then it is possible that bodies would undergo apparent vertical fall, because on a rotating Earth, a body before being released would be carried eastward by the Earth's rotation; and after being released this horizontal component of motion would be conserved. The body would also start moving downward; but this motion would not be a disturbance to the other one. Instead they would combine, by the principle of superposition, to produce the actually slanted path which would carry the body directly below the point of release, for example to the base of the tower, since the Earth's surface has also moved eastwards.

One last piece of reasoning was needed to complete Galileo's criticism. In that context, he could not simply assert the conservation of motion without justification. But he had one ready, which reflected the way he himself had arrived at the principle. The argument is in part an empirical one. Observation reveals that bodies accelerate as they move downwards, and slow down when they move upwards. Therefore, Galileo reasoned, if a body were moving along a path that was neither downwards nor upwards, its motion should be neutral, so to speak; its speed should remain constant in the absence of disturbances. But horizontal motion is an instance of motion which is neither upwards nor downwards. Therefore, bodies moving in a horizontal direction will conserve their speed of motion if left undisturbed.

So Galileo showed that the vertical fall objection is based on some untenable assumptions and is therefore groundless if vertical fall in its premises means apparent vertical fall, while the objection begs the question if the premises refer to actual vertical fall.

It should be noted, however, that none of this, by itself, supports the Earth's motion, let alone proves it; here we have simply the criticism or refutation of an argument, not a counter-argument justifying the opposite conclusion. There were many other mechanical objections to the Earth's motion, involving such things as the ship's mast experiment, east–west gunshots, north–south gunshots, etc.; these had to be similarly criticized, separately and piecemeal. Galileo did precisely that in the *Dialogue*.

And as we saw, eventually Galileo did formulate several positive arguments for the Earth's motion, some contributed by himself and some adapted from Copernicus: the simplicity argument for diurnal terrestrial rotation; the argument from the law of the periods of revolutions; the simplicity argument from the heliocentrism of planetary revolutions; the coherence argument from the explanation of planetary retrogressions; the argument from the annual motion of sunspots; and the argument based on the explanation of the tides. These positive arguments, together with the criticism of the objections, confirmed Copernicanism in the sense of rendering the Earth's motion more likely to be true than the Earth's rest. However, such a confirmation relied crucially on observational evidence with the telescope. And this leads us to a distinct and crucial methodological element of rationality.

Judgment and Telescopic Observation

Galileo's attitude toward the geokinetic idea before his telescopic discoveries is, in my opinion, a beautiful illustration of *judgment*. We are no longer dealing primarily with questions of reasoning, for it wasn't merely a matter of reasoning one's way out of the various astronomical objections. Nor was it a matter of criticism, for there was no question of his willingness and ability to challenge authority, as shown by the fact that he was engaged in a program of physical research which was undermining the foundations of Aristotelian physics. Rather, we are dealing with questions of proportion and balance.

In his early career, Galileo did indeed appreciate the novelty of Copernicus's argument, and he had begun to conceive ways of refuting the physical objections, and ways of providing mechanical evidence in favor of the Earth's motion. All that this meant was that physics and the criterion of explanatory coherence favored the geokinetic idea. But direct observation was still fully on the side of the geostatic view, and Galileo could not bring himself to any one-sided disregard of sense experience. That situation changed only with the telescope.

The telescope made possible the observation of phenomena that enabled answering the objection from the deception of the senses (together with the general observational argument for the geostatic view), the objection from the Earth–heaven dichotomy, and most of the specific astronomical objections. In regard to the latter, the planet Mars could now be seen to vary in apparent brightness and size as required by the hypothesis of the Earth's annual revolution. Similarly, the planet Venus exhibited the required phases.

The Earth–heaven dichotomy was undermined and the way paved for a unified view of heaven and Earth. The Moon's surface could be seen to be full of mountains and valleys like the Earth, and they could be seen to be nonluminous and to cast shadows from sunlight, very much as happens on the Earth. The Sun appeared to have on its surface dark spots that underwent changes similar to those of clouds on Earth. The planet Jupiter had four moons analogous to the one circling the Earth. And the phases of Venus also indicated that it was not composed out of a luminous aether, but out of some opaque nonluminous substance like the Earth.

Finally, the general observational argument for the geostatic view could now be criticized. One could say that, besides the direct experience of the unaided senses, the indirect observations made with the telescope should be taken into account. Since almost all indirect observation favored the Earth's motion, at the very least it was no longer true to say that observation unequivocally favored the geostatic system, and perhaps it was possible to say that it favored the Copernican system.

These discoveries may be said to have tipped the overall balance of evidence and argument in favor of the geokinetic and against the geostatic idea. Consequently, Galileo became increasingly outspoken about the issue, and in general an irreversible historical trend was produced which was to result in the eventual triumph of the geokinetic theory. However, the process was slow and gradual. The telescopic discoveries did not immediately decide the issue.

One reason was that at least one important astronomical objection could still not be answered—the argument from annual stellar parallax. Even the telescope did not reveal any yearly change in the apparent position of fixed stars. Galileo was correct in arguing that stellar distances are so immense that the parallax is very minute, and therefore more powerful instruments were needed to detect it.

Another reason why the telescope was not immediately decisive was that for some time there were proper concerns about the legitimacy, reliability, and practical operation of the instrument. Some questioned its legitimacy in principle, on the grounds that there was no place in scientific inquiry for instruments which make us see things that cannot be seen without them. Obviously, this objection could not be dismissed, as we can appreciate today if we compare the situation at that time with the more modern issue of whether psychedelic drugs put users in contact with a deeper level of reality, or merely make them see things that are not there. Others questioned the reliability of the telescope by pointing out that Galileo had not provided a scientific explanation of how and why the telescope worked. Moreover, all empirical checks involved terrestrial observation, and there was not even one instance of a test showing that it was truthful in the observation of phenomena in the heavens. Finally, the practical operation of the instrument required that one learn how to avoid aberrant and deviant observations stemming from impurities of the lenses, improper lens shape, and other features of poor design.[3]

A third reason why the telescopic discoveries were not decisive and did not provide a conclusive proof of the Earth's motion was the existence of biblical and other theological arguments against it. For example, as we have seen, one anti-Copernican passage in the Bible was taken to be Joshua 10:12–13, where God performs the miracle of stopping the Sun in its course in order to prevent it from setting at a place called Gibeon, and thus to give that region some extra daylight, needed by the Israelites to win a battle they were fighting. Understandably, it took some time for Galileo to come to terms with the scriptural objection by arguing that the Bible is not a scientific authority. Others, including the Church herself, of course, required an even longer time. Moreover, even though Galileo may have won all the arguments on this issue, he personally lost all the actual battles.

Finally, there was an epistemological issue connected with the objection from the deception of the senses which also required time for a full assimilation. The difficulty was that, although deception may be too strong a word, the Earth's motion was then and remains today a phenomenon which is not observable either with telescopes or by astronauts from outer space. Do such unobservable processes have any role in science, and if so what is their role? Can they be taken seriously as descriptions of physical reality, or can they only be regarded as useful fictions, useful, that is, for the calculation, computation, and prediction of other phenomena that are indeed observable? To admit unobservable entities in the scientific description of the world was a giant step for mankind, to be undertaken with great caution and circumspection.

Galileo's skepticism about Copernicanism before the telescope, and his caution about its observational status afterwards, attest to the importance of observation and judiciousness in the Copernican Revolution. He could not bring himself to take Copernicanism seriously, disregarding the pre-telescopic observational counter-evidence and exaggerating the role of the theoretical arguments in favor, until the telescope made possible a new kind of observation which could be judged to favor Copernicanism.

Fair-mindedness, Not Sophistry

Recall that in the winter of 1615–16 Galileo went to Rome to try to defend himself and Copernicanism, while the Inquisition was conducting an investigation started by some formal complaints against his views. From an intellectual and methodological point of view, his discussions there exhibited a memorable technique: before criticizing the anti-Copernican arguments, he made sure that their meaning was understood and their strength appreciated. Many scholars have misinterpreted this technique as the sophistical art of confusing and ridiculing opponents by first defending one thesis and then its opposite, and thus winning an argument even when logic and evidence are against you. I argued that such a technique is a sign that one is not demonizing one's opponents, but treating them as reasonable people; and it is a very effective manner of arguing, both to convince other people and to arrive at the truth.

This technique involves skills and qualities of mind that may be labeled open-mindedness and fair-mindedness. Generally speaking, Galileo's *Dialogue* is full of such open-minded and fair-minded argumentation. The best example is probably his discussion of the argument from the extruding power of whirling.

To recap, this objection argued that the Earth cannot rotate because if it did, bodies on its surface would have to move along curved paths following the Earth's circumference; and if they moved in this manner, then they would experience an extruding power away from the Earth's center, and so they would fly off toward the sky; but, obviously, this phenomenon is not observed. This argument was one of the mechanical objections to the Earth's motion, whose full answer required the discovery and understanding of the laws of circular motion and centrifugal force, toward which Galileo made some preliminary contributions.

To better explain Galileo's critique of this extrusion argument, we can refer to one of his own diagrams. In Figure 22, CEH represents a portion of the Earth's surface; A, the Earth's center; AC, a terrestrial radius; CD, a tangent. The hypothetical terrestrial rotation is taken to be in a counterclockwise direction (here, right to left). Galileo's refutation of the argument consists of three objections.

First, if a body were to be extruded from a rotating Earth, the extrusion would occur along the tangent (line CD) to the point of last contact with the terrestrial surface; the reason for this stems from the principle of conservation of motion (what we now call inertia). But, because of gravity, on a rotating Earth the body would still have a tendency to move downward along the

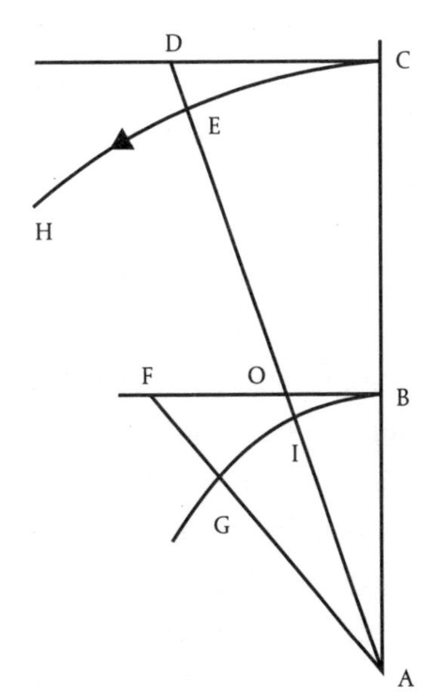

Figure 22. Extruding power of whirling on a rotating Earth

secant (DA) from the point of its position to the center of the Earth. So we need to do a comparison between these two tendencies; we can't consider just the centrifugal extrusion, as the anti-Copernican argument seems to be doing. The comparison shows that, in fact, the downward tendency happens to be greater than the extruding one.

In the second criticism, Galileo tries to show that the downward tendency not only happens to exceed the tangential one, but that it necessarily does so for mathematical reasons; that is, he argues that extrusion would be mathematically impossible on a rotating Earth. He tries to prove this mathematical impossibility on the basis of the geometry of the situation in the neighborhood of the point of contact between a circle and a tangent, and the behavior of the external segments (called exsecants, such as DE) of the secants (such as AD) drawn from the center of the circle to the tangent. The key point in this mathematical demonstration is that, as one gets closer and closer to the point of tangency (C), the exsecants (such as DE) get smaller and smaller at a much faster rate than the tangential segments (such as DC), and so the distances of fall required to prevent extrusion become infinitely smaller than the distances required to achieve extrusion.

Galileo's third objection concerns the point that in circular motion, the extruding tendency increases with the linear speed but decreases with the radius. Now, on a rotating Earth, the linear speed at the equator would be about 1,000 miles per hour, which sounds high in absolute terms; but such a speed is actually very small compared with the Earth's circumference, which is about 24,000 miles, or even with the Earth's radius, which is about 4,000 miles. In other words, the Earth's rotation is really very slow, in the sense that the Earth rotates just once in 24 hours. Thus, on a rotating Earth the extruding tendency would really be very small. This can be illustrated in the figure by imagining that that the Earth's radius were much smaller than it is (AB instead of AC), and that the linear speed were the same (BG = CE, both traversed in the same time); then the extruding tendency would be a function of the exsecant DE for the larger radius, and a function of exsecant FG for the smaller radius; and DE is smaller than FG. The anti-Copernican argument ignores this aspect of the situation, and so it assumes that on a rotating Earth extrusion would be more likely to happen than is the case.

This critique of the extrusion argument is an attempt to show that the argument is quantitatively invalid, based on an analysis of the mathematical physics of terrestrial extrusion. When so reconstructed, Galileo's refutation is essentially correct, but not completely. This is especially true for his second criticism, which tries to prove the mathematical impossibility of extrusion on a rotating Earth; this proof contains parts that are mathematically valid but are misapplied to the physical situation, and parts that would be physically applicable but are mathematically incorrect.[4]

However, the more important point in the present context is that before refuting the extrusion argument, Galileo amplifies and strengthens it in several ways. One is that he gives some examples and evidence to establish the reality of the extruding power of whirling, which is a crucial premise of the argument. For example, he mentions the experiment of tying a small pail of water at the end of a string, making a small hole in the bottom of the pail, whirling the pail in a vertical circle by the motion of one's hand, and observing water rushing out of the hole always in a direction away from one's hand. Galileo makes his intention clear in the *Dialogue*. Referring to the extrusion argument, he says that "I want to strengthen and tighten it further by showing even more sensibly how true it is that, when heavy bodies are rapidly turned around a motionless center, they acquire an impetus to move away from that center, even if they have a propensity to go toward it naturally."[5]

Even more striking is that Galileo makes an essential clarification. He points out that as usually formulated, the argument is improperly stated: its crucial step should be stated to say that if the Earth were rotating then there would now be no loose bodies on its surface, since they would have all been extruded long ago, rather than claiming that if the Earth were in rotation we would see bodies on its surface extruded off toward the sky.

Finally, Galileo makes a third fair-minded point by defending the extrusion argument from the unfair criticism that it commits the classic Aristotelian fallacy of *ignoratio elenchi*.[6] This would mean in effect that the argument alleges to be proving one conclusion (that the Earth is not in rotation), which is indeed controversial, but instead at best proves another (that the Earth did not recently begin to rotate), which nobody is claiming; in other words, the argument reaches an irrelevant conclusion—a proposition that is not being disputed. Such a criticism would be unfair because once it is made, it becomes immediately obvious how the original argument should be restated and was intended to be stated.

These improvements in the extrusion argument are so many and so important that, together with the fact that it is hard to find this particular argument in the texts of Aristotle or Ptolemy, they have led at least one critic to claim that the argument was largely invented by Galileo so that he could refute it![7] This claim is advanced as a criticism that exposes another one of the alleged sophistical tricks in Galileo's bag, that he was setting up a "straw man." But such criticism again inverts the truth, by attempting to portray negatively a technique that is actually sound and valuable.

Formulating arguments against your own position is a powerful method for the discovery of the truth. It is neither common nor easy, and certainly not for the faint-hearted; and it is usually a task to be left to one's opponents. But if one can pull it off, and if one can then undertake an evaluation and criticism of such counter-arguments, this procedure provides another indication of where the truth lies, as our own position is being made to overcome additional obstacles. Galileo may not have invented a novel argument, but there is little doubt that he greatly strengthened and amplified an existing one.

Another example of Galilean fair-mindedness concerns the anti-Copernican argument from the apparent position of fixed stars, which was based on the fact that no annual stellar parallax could be observed. We have seen on many occasions that Galileo could not really refute this argument for the simple reason that even the telescope did not reveal a stellar parallax. I pointed out that Galileo was explicit in the *Dialogue* that the parallax argument could not be refuted, and the main point of his discussion was to sketch a research program designed to measure the parallax if it really existed. But Galileo also did something else. He began his discussion by clearing up some common misconceptions about the kind of stellar changes that would be produced by the Earth's annual motion, in order to make sure that one understood the nature of the anti-Copernican argument.

One misunderstanding was that the elevation of the celestial pole above the horizon would change; this betrayed a failure to understand that if the Earth revolved around the Sun, the celestial pole would be defined by the terrestrial pole around which the Earth would rotate, and the elevation of the latter can change only by moving around the Earth's surface, and not by any motion of the whole Earth. Another confusion was that the annual change in the apparent position of fixed stars would be comparable to the large changes in stellar elevation above the horizon resulting by moving around the Earth's surface; this misunderstood the difference between moving on a curved surface (the Earth's) while measuring stellar elevations relative to that surface, and moving on a plane (the ecliptic) while measuring stellar elevations relative to that plane. A third misconception was that the annual changes for the fixed stars would be comparable to the large changes easily observable in the elevation of the Sun over the horizon, which generate the cycle of the seasons; such thinking failed to appreciate the difference between moving in an orbit around a body—the Earth around the Sun—and moving in an orbit far away from a body—the Earth's motion against the fixed stars.[8]

A Conceptual Framework

So far in this chapter, I have argued that nothing compares with the Copernican Revolution as a vivid illustration of the possibility and importance of criticism. The lesson here extends far beyond science: everyone may be mistaken, and everything is and ought to be open to criticism.

However, man is indeed, as Aristotle declared, a rational animal. Universal human beliefs are normally rational and reasonable, and such was certainly the geostatic belief before Copernicus and Galileo. Therefore, criticism would lack judgment if it did not recognize the importance of reasoning.

On the other hand, reasonableness and rationality are matters of degree, and they are contextual. What is reasonable under certain conditions need not always remain so. Still, a change will not occur arbitrarily, but rather only when a prevailing idea is shown to be less reasonable than a new idea. Again, judgment is required to ensure that reasonable ideas are not discarded arbitrarily, but only in the light of more reasonable ones.

These claims about the Copernican Revolution and Galileo's contributions can in turn be reinterpreted as conclusions about the nature of human rationality in general. For if we take these events as a defining instance of human rationality in science, then our account also enables us to see that and to see how human rationality involves three elements: criticism, reasoning, and judgment. These notions now require some clarification, definition, amplification, and systematization,

By reasoning, I mean a form of thinking consisting of the interrelating of thoughts in such a way as to make some thoughts dependent on others, and this interdependence can take the form of some thoughts being based on others or some thoughts following from others. Here we are defining reasoning in terms of thinking, and taking thinking as an undefined primitive notion.

"Criticism" can have several different meanings. In the narrow sense of negative evaluation, as used earlier in this chapter, we could speak of criticism as thinking that is aware of and open to the possibility of the falsehood or refutability of a proposition, including one's own view. We can give the label *principle of fallibility* to a key lesson of the Copernican Revolution, that everyone may be mistaken and everything is and ought to be open to criticism. Then, criticism in this narrow sense is thinking guided by or aware of the principle of fallibility.

But there is another meaning of *criticism*, which refers to self-awareness or self-reflection. In this sense, criticism is thinking that displays an awareness of what one is doing. When what we are doing is searching for the truth or trying to acquire knowledge, such criticism is equivalent to *methodological (or epistemological) reflection*: that is, thinking aimed at the formulation, analysis, or application of principles about the proper procedure to follow in the search for truth and the quest for knowledge.

A third common meaning of criticism is *evaluation*, in the general sense of either favorable and positive assessment or unfavorable and negative assessment. This is the meaning one has in mind when one speaks of art criticism or aesthetic criticism.

Now, going back to reasoning, a special case, *argument*, is reasoning that aims to justify a conclusion by supporting it with reasons and/or defending it from objections. This enables us to focus on another special case of crucial importance, *critical reasoning*, which I would define as reasoning aimed at, or consisting of, the interpretation, evaluation, or self-reflective formulation of arguments. Such self-reflection may involve not only the principle of fallibility, but also others, such as open-mindedness, fair-mindedness, rational-mindedness, and judicious-mindedness (or, more simply, judiciousness, or judgment).

These, in turn, may be defined as follows. Open-mindedness is the ability and willingness to know, understand, and learn from the arguments, evidence, and reasons against one's own views. Fair-mindedness is the ability and willingness to appreciate the strength of arguments and reasons against one's own view, even when one is attempting to criticize or refute them. Rational-mindedness is the ability and willingness to accept the views justified by the best arguments and strongest evidence; so defined, it should not be equated with rationality in general, but is rather a particular (although important) intellectual trait that is part of rationality.

Next, judiciousness is the willingness and ability to be impartial, balanced, and moderate; that is, to avoid one-sidedness (by properly taking into account all distinct aspects of an issue) and to avoid extremism (by properly taking into account the two opposite sides of any one aspect). This does not mean that anything goes, that one indiscriminately accepts all points of view as equally good, or that one does not accept any one point of view as better; nor does it mean that one mechanically splits the differences separating the several aspects and the opposite sides. Instead, the view which one accepts must be "properly" balanced and moderate.

Finally, *critical thinking* may then be considered a special case of critical reasoning that stresses such principles. That is, critical thinking is reasoning aimed at, or consisting of, the interpretation, evaluation, or self-reflective formulation of arguments, and guided by such ideals as the principles of fallibility, open-mindedness, fair-mindedness, rational-mindedness, and judiciousness.

Let's see how this conceptual framework corresponds to Galileo's contributions to the Copernican Revolution. Before the telescope, he was keenly aware of the strength of the anti-Copernican arguments based on the Earth–heaven dichotomy, the appearance of Venus, the apparent brightness and size of Mars, and the apparent position of fixed stars. This appreciation exemplifies his open-mindedness, fair-mindedness, and judiciousness. After the telescope, his critical reasoning about these arguments and the new evidence, together with his rational-mindedness, enabled him to re-assess and to seriously and actively pursue the Copernican theory.

Similarly, we have seen that the *Letter to Christina* provides a clear statement, an appreciative interpretation, and a nuanced evaluation of the scriptural argument against the Earth's motion. Galileo's essay should be seen as an instance of critical reasoning, not an abstract treatise on hermeneutics.

And Galileo's critique in the *Dialogue* of the anti-Copernican arguments likewise exemplifies these concepts. The discussion of the argument from the extruding power of whirling is first and foremost an instance of critical reasoning. It also carries open-mindedness to a new height, for it stresses an objection to Copernicanism that is more powerful than most of those advanced by the anti-Copernicans themselves. And it displays fair-mindedness by ensuring that this anti-Copernican argument is properly stated, before it is refuted. Nevertheless, it is ultimately invalid, and rational-mindedness dictates its rejection.

Finally, the criticism of the argument from vertical fall is obviously a piece of critical reasoning. The analysis leads not only to rejecting vertical fall as evidence against the Earth's motion (as demanded by rational-mindedness), but also to a better understanding of the objection than one finds in the proponents of geocentrism themselves (thus exemplifying open-mindedness and fair-mindedness).

These concepts and principles correspond not merely to Galileo's practice, but also to his reflections. In the Third Day of the *Dialogue*, in the context of the criticism of the empirical astronomical objections to the Earth's motion, he expresses amazement at "how in Aristarchus and Copernicus their [aprioristic] reason could have done so much violence to their senses, as to become, in opposition to the latter, mistress of their belief."[9] On the other hand, he confesses that in his own case, if telescopic observation "had not joined with reason, I suspect that I too would have been much more recalcitrant against the Copernican system than I have been."[10] These pronouncements express the importance of being judicious with regard to the contrast between theoretical speculation and sensory observation.

An eloquent statement of Galileo's open-mindedness is found in the Second Day of the *Dialogue*, in the course of the presentation of the many objections to Copernicanism. The Copernican character Salviati is keen to point this out to the Aristotelian Simplicio: "you will hear the followers of the new system produce against themselves observations, experiments, and reasons much stronger than those produced by Aristotle, Ptolemy, and other opponents of the same conclusions; you will thus establish for yourself that it is not through ignorance or lack of observation that they are induced to follow this opinion."[11]

And with regard to rational-mindedness, in a set of notes meant to provide replies to the objections in Bellarmine's letter to Foscarini, Galileo states:

> Not to believe that there is a demonstration of the earth's mobility until it is shown is very prudent, nor do we ask that anyone believe such a thing without a demonstration. On the contrary, we only seek that, for the advantage of the Holy Church, one examine with the utmost severity what the followers of this doctrine know and can advance, and that nothing be granted them unless the strength of their arguments greatly exceeds that of the reasons for the opposite side.[12]

On the principle of fair-mindedness, Galileo found the occasion to formulate it with words that are memorable for their clarity, elegance, and eloquence. The context was the complex and problematic one of the Inquisition proceedings of 1633. In the second deposition, he pleaded guilty to having, in the *Dialogue*, violated Bellarmine's warning not to defend the truth of Copernicanism. But he justified his violation as being the unintentional result of his wanting to critically examine the pro-Copernican arguments: "when one presents arguments for the opposite side with the intention of confuting them, they must be explained in the fairest way and not be made out of straw to the disadvantage of the opponent."[13]

The Copernican Revolution, then, required that the geokinetic hypothesis be justified not only with new theoretical arguments but also with new observational evidence; that the Earth's motion be not only constructively supported with theoretical arguments and observational evidence, but also critically defended from many powerful old and new objections; and that this defense include not only the destructive refutation but also the appreciative understanding of those objections in all their strength. One of Galileo's major accomplishments was not only to provide new evidence (from telescopic observation) supporting the Earth's motion, but also to show how the anti-Copernican objections could be refuted, and to elaborate their power before they were answered. In this sense, Galileo's contribution to the Copernican Revolution was rational-minded, open-minded, fair-minded, and judicious. He was and remains a model of critical reasoning and critical thinking. This is an everlasting and universally relevant intellectual lesson that can be learned from Galileo.

CHAPTER 11

SOME FINAL THOUGHTS

Appreciating Complexity and Simplicity

The Galileo affair displays not only conflicts between science and religion, but also harmonies between them. It embodies the complexity of the relationship between the two, as well as the way in which conflict and harmony may co-occur. It also captures the interplay between conservation and innovation, and between myths and facts. But I do not want to give the impression that the affair merely supports what has been called the complexity thesis.[1] Indeed, I believe that complexity *per se* is methodologically unsatisfactory, and that my account of the affair possesses a simplicity and elegance that transcend the conflicts and complexities.

To begin with, the Galileo affair should be explicitly separated into the *original* affair that climaxed in 1633, and the *subsequent* affair, which began after his condemnation and continues today. This distinction is important because there is no *a priori* reason why the relationship between science and religion that characterizes one of these two affairs should also characterize the other. And even if the same relationship were to hold in both controversies, the supporting evidence in the two cases would have to be different; it would be obviously invalid to argue that there was a conflict between science and religion in the original episode because there was such a conflict in the subsequent affair, or that there was no conflict in the subsequent affair because there was no such conflict in the original episode. Moreover, even if the relationship were conflictual in both cases, the particular form or direction of the conflict may very well be different (as we have seen).

Nor is there any *a priori* reason why the lessons about rationality, scientific method, and critical thinking that can be learned from the original and the subsequent affairs should be identical, though it may turn out that some of the lessons may be applicable to both. This is especially true for the application of lessons from the original affair to the subsequent controversy.

A further distinction can be made between the easily or directly discernable surface structure of a controversy and its deep structure of underlying characteristics, whose presence enables us to explain how and why the surface structure comes about. However, the deep structure does not *explain away* the surface structure; the reality of the latter does not disappear once it is explained in terms of the deep structure. We may also say that the deep structure causes the surface structure. However, here the cause-and-effect relationship is not deterministic or necessary, but a looser connection that is historical and contingent.

This distinction is important partly because one wants to know not only what "really" happened, but also "why" it happened. The deep structure also enables us not only to take into account the complexity of the situation, but also to transcend it into a greater simplicity.

Looking first at the surface structure of the original affair, I have claimed that it undeniably exhibits a conflict between science and religion. The key grounds for this conflict are the key events of Galileo's trial. Here, the conflict takes the form of religion vs. science, i.e., religion attacking science. The scientist Galileo was persecuted, tried, and condemned by institutions and officials of the Catholic religion.

The original affair has harmonious elements too, chiefly Galileo's own belief and cogent argument that there is no real conflict between Copernicanism and Scripture (an argument the Church would eventually accept). But such harmony does not destroy the conflict, which reappears in the historical context of the original affair as the Church holding the opposite view: that there really is a conflict between Copernicanism and Scripture.

At the deeper level, this conflict between Galilean science and Catholic religion can be understood as the result of a conflict between conservation and innovation, since many churchmen sided with Galileo, and some scientists sided with his opponents. There were even some key figures who individually embodied the conflict between conservation and innovation in their own thinking, attitudes, and actions. Pope Paul V was mostly a conservative, but in 1616 he did not declare Copernicanism a formal heresy and did not have Galileo tried and condemned; Cardinal Bellarmine, although primarily a conservative, was willing to tolerate Copernicanism as a hypothesis to explain observed facts and to make calculations and predictions; and Pope Urban VIII had innovative inclinations that encouraged Galileo to write the *Dialogue*, though eventually his conservatism prevailed. The science–religion conflict resulted from the conservation–innovation conflict in the sense that the conservatives and conservative attitudes happened to prevail over the innovators and innovative attitudes in Galileo's trial.

The surface structure of the subsequent affair can also be seen to consist of a conflict between science and religion, but this time taking the form of science vs. religion. For the past four centuries, the Catholic Church has been under fire from scientists and alleged representatives of the scientific attitude on account of its treatment of Galileo. Evidence for such a conflict is found in the writings of Milton, Voltaire, Libri, and Einstein, but these are merely the tip of an iceberg of anti-clerical and pro-Galilean criticism. The other main body of evidence consists of various apologetic attempts to defend the Church and blame Galileo, such as the examples mentioned earlier of Ross, Mallet du Pan, Brewster, Duhem, and Feyerabend; these examples are also the tip of an iceberg, although much smaller than the anti-clerical one. An additional piece of evidence for the conflict is the sequence of actions taken by the Church to retract or undo her condemnation of Copernicanism, of Galileo, and of the principle limiting the scientific authority of Scripture.

As for the deep structure of this subsequent conflict between science and religion, I have argued that it is to be found in the conflict between cultural myths and documented facts. For if we dig under the surface of the anti-clerical criticism and the pro-clerical apologetics, we find the phenomenon of myth-making and mythologizing, i.e., the rise, evolution, and fall of cultural myths in light of available or ascertainable facts. Accordingly, one would explore whether both sides engaged in their share of exaggerations, distortions, propaganda, and rhetoric; whether such literally false beliefs (of both types) are really impossible to discard; and whether they perform a necessary and useful social function.

So, to summarize, in my view, there is a religion vs. science conflict in the original Galileo affair and a reverse conflict in the subsequent Galileo affair, but both occur at the level of the surface structure. At a deeper level they result from conflicts of things other than science and religion: the dialectic between conservation and innovation in the case of the original affair, and the dialectic of myths and facts for the subsequent conflict. This is meant to explain, but not explain away, the conflict between science and religion in the Galileo affair.

On this last point, an analogy may be helpful. Physics explains the phenomenon of heat based on the motion and kinetic energy of the molecules that make up physical bodies. The kinetic theory enables us to understand many observed facts about heat and make additional predictions of less easily observable phenomena. But this does not undermine the reality of heat. Similarly, the reality of the science-and-religion conflict in the Galileo affairs is not undermined by the underlying conflicts between conservation and innovation and between myths and facts from which they arise.

This framework is, I think, complex enough to enable us to take into account the complications of the phenomenon we are dealing with, but also elegant enough to give it a kind of simplicity.

An Elegant Symmetry

There is an uncanny and elegant symmetry between some key elements of the original controversy and the subsequent one. It involves intellectual issues, partly at the level of interpretation or understanding, and partly at the level of evaluation or action; and it utilizes the lessons we derived from Galileo concerning rationality, scientific method, and critical thinking.

In this book, I have been arguing that an important aspect of the Copernican Revolution was the defense of the geokinetic hypothesis from a host of objections based on astronomical observation, Aristotelian physics, scriptural passages, and traditional epistemology. A major contribution to this defense was provided by Galileo. He answered the observational astronomical objections once the telescope revealed new celestial phenomena and revolutionized astronomical observation by making it instrument-based. He answered the scriptural objections by arguing that Scripture is not a scientific authority, and so scriptural passages should not be used to invalidate astronomical claims that are proved or provable. He answered the mechanical objections by articulating a new physics centered on the principles of conservation, composition, and relativity of motion. More generally, Galileo's key contribution to the Copernican Revolution was to elaborate a defense of Copernicanism that stressed reasoning and argumentation judiciously guided by the ideals of fallibility, open-mindedness, and fair-mindedness.

Despite Galileo's prudence and indirectness, and the support from many churchmen, his defense of Copernicanism was hindered by key officials and institutions of the Catholic Church. In fact, the trial of Galileo can be interpreted as a series of ecclesiastic attempts to stop him from such a reasoned defense of Copernicanism. In 1616, the Index decreed that the geokinetic doctrine was contrary to Scripture, and this decree amounted to a general prohibition on defending Copernicanism from scriptural objections. At the same time, Cardinal Bellarmine officially warned Galileo to cease holding and defending the truth of the Earth's motion, and this warning amounted to a personal prohibition on supporting or defending Copernicanism, except as a hypothesis to save the appearances and make calculations and predictions. In 1633, the Inquisition condemned Galileo as a suspected heretic, and this sentence amounted to condemning him for indirectly defending Copernicanism as probable in the *Dialogue* of 1632; for this book was primarily a critical discussion (i.e., an exercise in critical reasoning), examining the arguments on both sides, showing that the Copernican arguments were stronger than the geostatic ones, implying that the geokinetic hypothesis was probably true, and thus defending Copernicanism only indirectly and implicitly.

These condemnations, which represent the two principal phases of Galileo's trial (the original Galileo affair), in turn generated a much more protracted, complex, and controversial cause célèbre that continues to our own day. The subsequent Galileo affair is an intricate web of historical after-effects, reflective commentaries, and critical issues. However, a key interpretive idea for making sense of it is to focus on the many criticisms of Galileo's defense of Copernicanism (i.e., the many apologias of his condemnation) that have been advanced by his critics and on the various replies and counter-arguments put forth by his defenders. Such an interpretation then readily enables one to adopt one's own evaluative position about such anti-Galilean criticisms.

Note that this key interpretive idea corresponds to Galilean rational-mindedness: it focuses on the arguments of both sides and on the key claim which, as a result of such argumentation, is affirmed by one side and denied by the other. This key claim is the proposition that the Inquisition's condemnation of Galileo in 1633 was right. This is then taken to be the main issue of the subsequent controversy.

Next, in accordance with Galileo's ideal of open-mindedness, one focuses on the subsequent arguments trying to justify his condemnation and defend the Church (i.e., the anti-Galilean arguments), to see whether they have any validity.

For a while, various questions were raised about the physical truth of the Earth's motion; but gradually, the historical development of science established incontrovertibly that Galileo had been right on this issue. As this realization was emerging, questions began to be raised about whether his supporting reasons, arguments, and evidence had been correct; this is instructive, but Galileo's reasoning can be defended from this criticism. For some time, Galileo was also criticized for his hermeneutical principle that Scripture is not a scientific authority; but historical and cultural development also vindicated him in this regard—at least this is what happened from the point of view of what has become the official position of the modern Catholic Church. However, on the hermeneutical issue too, it is important to check the correctness of his arguments justifying that Scripture is not scientific authority; although this Galilean reasoning has been the target of many objections, I believe it can be defended from them.

As it became increasingly clear that Galileo could not be validly accused of being a bad scientist, a bad theologian, or a bad logician, he started being blamed for other reasons. Some authors began to stress the legal aspect of the trial, charging that he had been guilty of disobeying the Church's admonition regarding Copernicanism. But the content of this admonition is ambiguous. If the admonition is interpreted to be a prohibition on any kind of defense of Copernicanism (in accordance with Commissary Seghizzi's precept), the legitimacy and perhaps even the occurrence of such a special injunction is undermined by the record of the trial proceedings, which was first published in 1867–78; whereas if the admonition is taken to be a prohibition on defending the truth of Copernicanism (in accordance with Cardinal Bellarmine's testimony), then the issue reduces primarily to the question whether Galileo's defense was scientifically and logically fair and valid, and secondarily to the question whether even such a milder prohibition was legitimate.

Whether or not Galileo can be defended from such scientific, theological, logical, and legal criticisms, he can be and has been the target of epistemological criticism. In 1908, Duhem tried to blame him for his epistemological realism and argued that the condemnation would have been avoided if epistemological instrumentalism had prevailed. I believe Galileo can be defended from the charge that he was a bad epistemologist.

Next, there is the issue of whether Galileo is to be credited or blamed for helping us understand that science and religion are in conflict or that they are in harmony. The resolution of this issue requires that we reflect on three things: that the trial embodied a minimal but irreducible historical conflict between those who affirmed and those who denied that Copernicanism contradicted Scripture; that the trial epitomized more the conflict between conservation and innovation than the conflict between science and religion; and that because of how the trial was subsequently perceived, the conflict between science and religion is indeed an essential feature of the subsequent affair.

Finally, there is the current spectacle of the Galileo affair. One the one hand, we see the phenomenon of the rehabilitation movement within the

Catholic Church. On the other hand, one can witness the rise of socially oriented critiques of Galileo by leftist sympathizers and self-styled progressives. And we also observe the conflict between these two points of view, as well as the irony of the switching of sides.

In other words, the Copernican Revolution required that the geokinetic hypothesis be not only supported with new arguments and evidence, but also defended from many powerful old and new objections. This defense in turn required not only the destructive refutation but also the appreciative understanding of those objections in all their strength. One of Galileo's major accomplishments was not only to provide new evidence and arguments supporting the Earth's motion, but also to show how those objections could be refuted, and to elaborate their power before they were answered. In this sense, as we saw earlier, Galileo's defense of Copernicanism was reasoned, critical, open-minded, fair-minded, and judicious.

Now, an essential thread of the subsequent Galileo affair has been the emergence of many anti-Galilean criticisms, from a variety of viewpoints, including scientific, theological, philosophical, legal, and social. Such criticisms arise naturally and legitimately, but Galileo has been, or can be, effectively defended from them. The proper and effective way of defending him is by ensuring that we know and understand the anti-Galilean criticisms, and that we appreciate their strength before refuting them, thus modeling our own approach to the defense of Galileo on his approach to the defense of Copernicanism. Thus, today, in the context of the Galileo affair and the controversies over the relationship between science and religion and between institutional authority and individual freedom, *the proper defense of Galileo should have the reasoned, critical, open-minded, fair-minded, and judicious character present in his own defense of Copernicus.* This overarching thesis may be regarded as a critical interpretation of the Galileo affair, understood in terms of an original and a subsequent phase, and interpreted in terms of a stress on reasoning, argument, and evidence.

Such a view focuses on the reasons for accepting the ancient geostatic world view, for taking Copernicanism seriously, and for rejecting the physical truth of the geokinetic hypothesis; the reasons that led Galileo to pursue it, indirectly at first and directly later, so as to come to think that it was much more probable than the geostatic thesis; the reasons that led the Inquisition to prosecute and persecute Galileo; the reasons why his condemnation could be claimed and has been claimed to be right (first from the astronomical and hermeneutical points of view, then from the logical, legal, and epistemological points of view, and finally from the points of view of the cultural interaction between science and religion and between science and society); and the reasons why such anti-Galilean criticisms can be and have been criticized. In short, the principal interpretive thesis is that the reasoned defense of Copernicanism was Galileo's chief offense in the trial and his key contribution to the Copernican Revolution, while the defense of Galileo from various attempts to justify his condemnation is the most central issue in the subsequent affair.

This thesis also possesses a critical, evaluative, or normative dimension, for it amounts to saying that Galileo's reasoned and critical approach to defending Copernicanism is instructive and suggests the proper way in which Galileo himself should be defended. Besides trying to understand and interpret what really happened in the original Galileo affair, we should also try to assess and evaluate what in the current controversy is right or wrong from various nuanced points of view. When we do this, we discover that just as Galileo's defense of Copernicanism owed its success to its being reasoned, critical, and fair-minded, so our defense of Galileo can succeed if it possesses these same qualities.

In the light of this, my talk of "defending" Galileo should not be misunderstood. To defend Galileo does not mean to show that he was completely or perfectly right; it only means to show that he was essentially right, or more nearly correct than not. The defense of Galileo is not an attempt to show that criticisms of him are without foundation; on the contrary, the defense cannot even get started unless one first knows and understands that there are reasons for attributing to him various errors or improprieties; in such a context one tries to show that such anti-Galilean arguments are ultimately invalid, or at least weaker than the pro-Galilean ones. Defending Galileo is not meant to be a one-sided exercise pointing out only his merits and virtues, but rather a comparative analysis also taking into account his faults. Nor is the defense of Galileo a hagiographic exercise exaggerating the number or importance of his scientific, intellectual, and cultural achievements. My overarching thesis has, after all, a historical or interpretive component, besides the philosophical or evaluative one, and the main thrust of the interpretive component is the historical reality of the anti-Galilean criticisms.

Such a defense of Galileo should be interpreted as an exercise in critical reasoning, just as his defense of Copernicanism was. It follows, then, that the proposition that Galileo's defense of Copernicanism was wrong (i.e., that the Inquisition's condemnation of Galileo was right) is almost as false and untenable as the proposition that the Earth stands still at the center of the universe. On the other hand, the arguments purporting to justify various Galilean errors or improprieties are in appearance almost as plausible as the anti-Copernican arguments seemed to be in the sixteenth century. But ultimately the anti-Galilean arguments can be shown to be almost as weak and invalid as the anti-Copernican arguments were shown to be by Galileo. And here lies the uncanny and elegant symmetry between the two affairs.

BIBLIOGRAPHY

This list provides full bibliographical information for all the works that are cited in abbreviated form in the Notes. The list also includes a few works that are generally valuable, although they are not specifically cited in the Notes. Some of the works in both of these categories are recommended for further reading, and such a recommendation is indicated by means of an asterisk (*).

Abbott, Alison (2018). "Discovery of Galileo's Long-lost Letter Shows He Edited His Heretical Ideas to Fool the Inquisition." *Nature* 561(7724):441–2.

Acloque, Paul (1982). "L'Histoire des Expériences pour la Mise en Évidence du Mouvement de la Terre." *Cahiers d'Histoire et de Philosophie des Sciences*, new series, no. 4, 1–141.

Anonymous (2018). "Discovery of Galileo's Long-lost Letter Highlights the Value of Physical Repositories." *Nature* 561(7724):434.

Beltrán Marí, Antonio (1998). " 'Una Reflexión Serena y Objectiva'." *Arbor* 160(629):69–108.

*Beltrán Marí, Antonio (2006). *Talento y poder*. Pamplona: Laetoli.

Benedict XVI, Pope (2006). "Faith, Reason, and the University." Lecture delivered at the University of Regensburg, September 12. Available at http://w2.vatican.va/content/benedict-xvi/en/speeches/2006/september/documents/hf_ben-xvi_spe_20060912_university-regensburg.html; accessed June 28, 2017.

Benítez, Hermes H. (1999). *Ensayos sobre Ciencia y Religión*. Santiago: Bravo y Allende.

Beretta, Francesco (1998). *Galilée devant le Tribumal de l'Inquisition*. Doctoral Dissertation, Faculty of Theology, University of Fribourg, Switzerland.

*Beretta, Francesco (1999). "Le Procès de Galilée et les Archives du Saint-Office." *Revue des Sciences Philosophiques et Théologiques* 83:441–90.

Berggren, Lars, and Lennart Sjöstedt (1996). *L'ombra dei grandi*. Rome: Artemide Edizioni.

*Biagioli, Mario (1993). *Galileo Courtier*. Chicago: University of Chicago Press.

*Biagioli, Mario (2006). *Galileo's Instruments of Credit*. Chicago: University of Chicago Press.

*Blackwell, Richard J. (1991). *Galileo, Bellarmine, and the Bible*. Notre Dame: University of Notre Dame Press.

*Blackwell, Richard J. (2006). *Behind the Scenes at Galileo's Trial*. Notre Dame: University of Notre Dame Press.

Boaga, Emanuele (1990). "Annotazioni e documenti sulla vita e sulle opere di Paolo Antonio Foscarini teologo 'copernicano' (1562c.–1616)." *Carmelus* 37:173–216.

Brecht, Bertolt (1966). *Galileo*, trans. Charles Laughton, ed. Eric Bentley. New York: Grove Press.

Brewster, David (1841). *The Martyrs of Science, or the Lives of Galileo, Tycho Brahe, and Kepler*. London: John Murray.

*Brooke, John H. (1991). *Science and Religion*. Cambridge, UK: Cambridge University Press.

*Brooke, John, and Geoffrey Cantor (1998). *Reconstructing Nature*. Edinburgh: T & T Clark.

Brown, Harold I. (1979). *Perception, Theory and Commitment*. Chicago: University of Chicago Press.

*Bucciantini, Massimo (1995). *Contro Galileo*. Florence: Olschki.

*Bucciantini, Massimo (2003). *Galileo e Keplero*. Turin: Einaudi.

Bucciantini, Massimo, M. Camerota, and F. Giudice, Eds. (2011). *Il caso Galileo*. Florence: Olschki.

*Bucciantini, Masimo, M. Camerota, and F. Giudice (2015). *Galileo's Telescope*, trans. Catherine Bolton. Cambridge, MA: Harvard University Press.

*Butterfield, Herbert (1949). *The Origins of Modern Science, 1300–1800*. London: Bell.

*Camerota, Michele (2004). *Galileo Galilei e la cultura scientifica nell'età della Controriforma*. Rome: Salerno Editrice.

Camerota, Michele, Franco Giudice, and Salvatore Ricciardo (2018). "The Reappearance of Galileo's Original Letter to Benedetto Castelli." *Notes and Records of the Royal Society*. Doi:10.1098/rsnr.2018.0053. Published online October 25; available at http://rsnr.royalsocietypublishing.org/.

Campanella, Tommaso (1622). *Apologia pro Galileo*. Frankfurt.

Campanella, Tommaso (1994). *A Defense of Galileo*, trans. and ed. R. J. Blackwell. Notre Dame: University of Notre Dame Press.

Carvalho Neto, Paulo de (1971). *The Concept of Folklore*, trans. J. M. P. Wilson. Coral Gables, FL: University of Miami Press.

*Clavelin, Maurice (1974). *The Natural Philosophy of Galileo*, trans. A. J. Pomerans. Cambridge, MA: M.I.T. Press.

*Cohen, I. Bernard (1960). *The Birth of a New Physics*. Garden City, NY: Doubleday.

Copernicus, Nicolaus (1992). *On the Revolutions*, trans. and ed. E. Rosen. Baltimore: Johns Hopkins University Press.

*Dawes, Gregory W. (2016). *Galileo and the Conflict between Religion and Science*. London: Routledge.

*DiCanzio, Albert (1996). *Galileo: His Science and His Significance for the Future of Man*. Portsmouth, NH: ADASI.

Drake, Stillman (1957). *Discoveries and Opinions of Galileo*. Garden City, NY: Doubleday.

*Drake, Stillman (1978). *Galileo at Work*. Chicago: University of Chicago Press.

Drake, Stillman (1983). *Telescopes, Tides & Tactics*. Chicago: University of Chicago Press.

Drake, Stillman (1986). "Reexamining Galileo's *Dialogue*." In *Reinterpreting Galileo*, ed. William A. Wallace, pp. 155–75. Washington, DC: Catholic University of America Press.

Draper, John W. (1874). *History of the Conflict between Religion and Science.* New York: Appleton.

Dreyer, J. J. E. (1953). *A History of Astronomy from Thales to Kepler.* New York: Dover.

Duhem, Pierre (1908). *SOZEIN TA PHAINOMENA: Essai sur la notion de theorie physique de Platon à Galilée.* Paris: Hermann.

Duhem, Pierre (1969). *To Save the Phenomena: An Essay on the Idea of Physical Theory from Plato to Galileo*, trans. E. Doland and C. Maschler. Chicago: University of Chicago Press.

Einstein, Albert (1953). "Foreword." In Galilei (1953), pp. vi–xx.

Einstein, Albert (1954). *Ideas and Opinions*, trans. and ed. Sonja Bargmann. New York: Crown Publishers.

*Fantoli, Annibale (2003). *Galileo: For Copernicanism and for the Church*, 3rd edn., trans. G. V. Coyne. Vatican City: Vatican Observatory Publications.

*Fantoli, Annibale (2012). *The Case of Galileo*, trans. G. V. Coyne. Notre Dame: University of Notre Dame.

*Feldhay, Rivka (1995). *Galileo and the Church.* Cambridge, UK: Cambridge University Press.

Feyerabend, Paul K. (1975). *Against Method.* London: NLB.

*Feyerabend, Paul K. (1985). "Galileo and the Tyranny of Truth." In *The Galileo Affair: A Meeting of Faith and Science*, ed. George V. Coyne, M. Heller, and J. Zycínski, pp. 155–66. Vatican City: Specola Vaticana.

Feyerabend, Paul K. (1988). *Against Method*, rev. edn. London: Verso.

*Feyerabend, Paul K. (1993). *Against Method*, 3rd edn. London: Verso.

*Finocchiaro, Maurice A. (1980). *Galileo and the Art of Reasoning: Rhetorical Foundations of Logic and Scientific Method.* Boston Studies in the Philosophy of Science, vol. 61. Dordrecht: Reidel.

Finocchiaro, Maurice A. (1985). "Wisan on Galileo and the Art of Reasoning." *Annals of Science* 42:613–16.

*Finocchiaro, Maurice A., trans. and Ed. (1989). *The Galileo Affair: A Documentary History.* Berkeley: University of California Press.

*Finocchiaro, Maurice A., trans. and Ed. (1997). *Galileo on the World Systems: A New Abridged Translation and Guide.* Berkeley: University of California Press.

Finocchiaro, Maurice A. (2002a). "Galileo as a 'Bad Theologian': A Formative Myth about Galileo's Trial." *Studies in History and Philosophy of Science* 33:753–91.

*Finocchiaro, Maurice A. (2002b). "Philosophy versus Religion and Science versus Religion: The Trials of Bruno and Galileo." In *Giordano Bruno: Philosopher of the Renaissance*, ed. Hilary Gatti, pp. 51–96. Aldershot: Ashgate.

*Finocchiaro, Maurice A. (2005). *Retrying Galileo, 1633–1992.* Berkeley: University of California Press.

*Finocchiaro, Maurice A., trans. and Ed. (2008). *The Essential Galileo.* Indianapolis: Hackett.

Finocchiaro, Maurice A. (2009). "Myth 8: That Galileo Was Imprisoned and Tortured for Advocating Copernicanism." In Numbers (2009), pp. 68–78, 249–52.

*Finocchiaro, Maurice A. (2010). *Defending Copernicus and Galileo: Critical Reasoning in the Two Affairs.* Boston Studies in the Philosophy of Science, vol. 281. Dordrecht: Springer.

Finocchiaro, Maurice A. (2011). "Fair-mindedness vs. Sophistry in the Galileo Affair: Two Controversies for the Price of One." In *Controversies within the Scientific Revolution,* ed. Marcelo Dascal and Victor Boantza, pp. 53–73. Amsterdam: John Benjamins.

Finocchiaro, Maurice. A. (2013). "Galileo under Fire and under Patronage." In *Ideas under Fire,* ed. Jonathan Lavery, Louis Groarke, and William Sweet, pp. 123–43. Lanham: Rowman & Littlefield.

*Finocchiaro, Maurice A. (2014a). *The Routledge Guidebook to Galileo's Dialogue.* London: Routledge.

*Finocchiaro, Maurice A., trans. and Ed. (2014b). *The Trial of Galileo: Essential Documents.* Indianapolis: Hackett.

Finocchiaro, Maurice A. (2016). "Galileo's First Confrontation with the Inquisition (1616): Four Orders and Three Issues." *Galilaeana* 13:29–60.

Finocchiaro, Maurice A. (2018). "Authenticity vs. Accuracy vs. Legitimacy: Pagano on the Inquisition's 1616 Orders to Galileo." In *Incorrupta Monumenta Ecclesiam Defendunt: Studi offerti a mons. Sergio Pagano, prefetto dell'Archivio Segreto Vaticano,* 3 vols., ed. Andreas Gottsmann, Pierantonio Piatti, and Andreas E. Rehberg, vol. 3, pp. 183–200. Vatican City: Archivio Segreto Vaticano.

Foscarini, Paolo A. (1615). *Lettera sopra l'opinione de' pittagorici e del Copernico della mobilità della Terra e stabilità del Sole e del nuovo pittagorico sistema del mondo.* Naples.

Foscarini, Paolo A. (1991). "A Letter . . . Concerning the Opinion of the Pythagoreans and Copernicans About the Mobility of the Earth and the Stability of the Sun and the New Pythagorean System of the World . . ." Trans. R. J. Blackwell. In Blackwell (1991), pp. 217–51.

Galilei, Galileo (1890–1909). *Le Opere di Galileo Galilei,* 20 vols. National Edition by A. Favaro et al. Florence: Barbèra. Reprinted in (1929–39) and (1968).

*Galilei, Galileo (1953). *Dialogue Concerning the Two Chief World Systems, Ptolemaic and Copernican,* trans. and ed. Stillman Drake. Berkeley: University of California Press. Second edition (1967).

Galilei, Galileo (1960). *On Motion and on Mechanics,* trans. and ed. I. E. Drabkin and S. Drake. Madison: University of Wisconsin Press.

Galilei, Galileo (1982). *Dialog über die beiden hauptsächlichsten Weltsysteme, das ptolemäische und das kopernikanische,* trans. E. Strauss (1891), ed. R. Sexl and K. von Meyenn. Stuttgart: Teubner.

*Galilei, Galileo (1997). *Galileo on the World Systems: A New Abridged Translation and Guide,* trans. and ed. M. A. Finocchiaro. Berkeley: University of California Press.

Galilei, Galileo (2002). *Dialogues Concerning Two New Sciences,* ed. Stephen Hawking. Philadelphia: Running Press.

*Galilei, Galileo (2008). *The Essential Galileo,* ed. and trans. M. A. Finocchiaro. Indianapolis: Hackett.

Galilei, Galileo (2012). *Selected Writings,* trans. and ed. W. R. Shea and Mark Davie. Oxford: Oxford University Press.

Galluzzi, Paolo (1977). "Galileo contro Copernico." *Annali dell'Istituto e Museo di storia della scienza* 2:87–148.

Galluzzi, Paolo (2000). "Gassendi and l'*Affaire Galilée* of the Laws of Motion." *Science in Context* 13:509–45.

Genovesi, Enrico (1966). *Processi contro Galileo.* Milan: Casa Editrice Ceschina.

*Gingerich, Owen (1982). "The Galileo Affair." *Scientific American*, August, 132–43.

Guerrini, Luigi (2009). *Galileo e la polemica anticopernicana a Firenze.* Florence: Polistampa.

Harris, W. H., and J. S. Levey, Eds. (1975). *The New Columbia Encyclopedia.* New York: Columbia University Press.

Hawking, Stephen W. (1988). *A Brief History of Time.* New York: Bantam Books.

Hawking, Stephen W. (2002). "Galileo Galilei (1564–1642): His Life and Work." In Galilei (2002), pp. xi–xvii.

*Heilbron, J. L. (1999). *The Sun in the Church.* Cambridge, MA: Harvard University Press.

Heilbron, J. L. (2005). "Censorship of Astronomy in Italy after Galileo." In McMullin (2005), pp. 279–322.

*Heilbron, J. L. (2010). *Galileo.* Oxford: Oxford University Press.

*Henry, John (1997). *The Scientific Revolution and the Origins of Modern Science.* New York: St. Martin's Press.

*Henry, John (2011). "Galileo and the Scientific Revolution." *Galilaeana* 8:3–36.

Hill, David K. (1984). "The Projection Argument in Galileo and Copernicus." *Annals of Science* 41:109–33.

Howell, Kenneth J. (2002). *God's Two Books.* Notre Dame: University of Notre Dame Press.

Hume, David (1851–60). *The History of England from the Invasion of Julius Caesar to the Abdication of James the Second, 1688*, 6 vols. New York: Harper.

Hume, David (1948). *Dialogues Concerning Natural Religion*, ed. H. D. Aiken. New York: Hafner.

John Paul II, Pope (1979). "Deep Harmony Which Unites the Truths of Science with the Truths of Faith." *L'Osservatore Romano*, Weekly Edition in English, November 26, pp. 9–10.

John Paul II, Pope (1992). "Faith Can Never Conflict with Reason." *L'Osservatore Romano*, Weekly Edition in English, November 4, pp. 1–2.

Kant, Immanuel (1965). *Critique of Pure Reason*, trans. N. Kemp Smith. New York: St. Martin's Press.

Kelly, Henry Ansgar (2015). "Judicial Torture in Canon Law and Church Tribunals." *Catholic Historical Review* 101:754–93.

Kelly, Henry Ansgar (2016). "Galileo's Non-Trial (1616), Pre-Trial (1632–1633), and Trial (May 10, 1633)." *Church History* 85:724–61.

Koestler, Arthur (1959). *The Sleepwalkers.* New York: Macmillan.

Koestler, Arthur (1964). "The Greatest Scandal in Christendom." *Observer* (London), February 2, pp. 21, 29.

Koyré, Alexandre (1955). *A Documentary History of the Problem of Fall from Kepler to Newton.* In *Transactions of the American Philosophical Society*, new series, vol. 45, part 4, pp. 329–95. Philadelphia: American Philosophical Society.

*Kuhn, Thomas S. (1957). *The Copernican Revolution*. Cambridge, MA: Harvard University Press.

*Kuhn, Thomas S. (1970). *The Structure of Scientific Revolutions*, second edn. Chicago: University of Chicago Press.

Kuhn, Thomas S. (1977). *The Essential Tension*. Chicago: University of Chicago Press.

Leo XIII, Pope (1893). *Providentissimus Deus*. In *The Papal Encyclicals, 1740–1981*, 5 vols, ed. Claudia Carlen, vol. 2, pp. 325–39. Wilmington, NC: McGrath (1981).

Lerner, Michel-Pierre, trans. and ed. (2001). *Tommaso Campanella, Apologia pro Galileo*. Paris: Les Belles Lettres.

Lessl, Thomas S. (1999). "The Galileo Legend as Scientific Folklore." *Quarterly Journal of Speech* 85:146–68.

Libri, Guglielmo (1841). *Essai sur la Vie et les Travaux de Galilée*. Paris.

Lightman, Bernard (2001). "Victorian Sciences and Religions." In *Science in Theistic Contexts*, ed. John H. Brooke, M. J. Osler, and J. M. van der Meer, pp. 343–66. *Osiris*, second series, vol. 16. Chicago: University of Chicago Press.

Lindberg, David C. (1992). *The Beginnings of Western Science*. Chicago: University of Chicago Press.

*Lindberg, David C. (2003). "Galileo, the Church, and the Cosmos." In Lindberg and Numbers (2003), pp. 33–60.

Lindberg, David C., and R. L. Numbers, Eds. (1986). *God and Nature*. Berkeley: University of California Press.

Lindberg, David C., and R. L. Numbers (1987). "Beyond War and Peace: A Reappraisal of the Encounter between Christianity and Science." *Perspectives on Science and Christian Faith* 39:140–9.

Lindberg, David C., and R. L. Numbers, Eds. (2003). *When Science & Christianity Meet*. Chicago: University of Chicago Press.

Livingstone, David N. (2007). "Science, Religion, and the Cartographies of Complexity." *Historically Speaking: Bulletin of the Historical Society* 8(5):15–16.

MacLachlan, James H. (1990). "Drake Against the Philosophers." In *Nature, Experiment, and the Sciences*, ed. Trevor H. Levere and W. R. Shea, pp. 123–44. Dordrecht: Kluwer.

McMullin, Ernan, Ed. (2005). *The Church and Galileo*. Notre Dame: University of Notre Dame Press.

Mallet du Pan, Jacques (1784). "Mensognes Imprimées au Sujet de la Persécution de Galilée." *Mercure de France*, July 17, pp. 121–30.

Masini, Eliseo (1621). *Sacro arsenale overo prattica dell'officio della Santa Inquisizione*. Genoa: Giuseppe Pavoni.

Mayaud, Pierre-Noël, S. J. (1997). *La Condamnation des Livres Coperniciens et sa Révocation*. Rome: Editrice Pontificia Università Gregoriana.

Mayer, Thomas F., trans. and Ed. (2012). *The Trial of Galileo 1612–1633*. Toronto: University of Toronto Press.

Mayer, Thomas F. (2013). *The Roman Inquisition: A Papal Bureaucracy and Its Laws in the Age of Galileo*. Philadelphia: University of Pennsylvania Press.

Mayer, Thomas F. (2014). *The Roman Inquisition on the Stage of Italy, c. 1590–1640*. Philadelphia: University of Pennsylvania Press.

Mayer, Thomas F. (2015). *The Roman Inquisition: Trying Galileo*. Philadelphia: University of Pennsylvania Press.

Miller, David M. (2008). "The Thirty Years War and the Galileo Affair." *History of Science* 46:49–74.

Milton, John (1644). *Areopagitica*. In Milton (1953–82), vol. 2, pp. 485–570.

Milton, John (1953–82). *Complete Prose Works*, 8 vols. New Haven, CT: Yale University Press.

Moore, James R. (1979). *The Post-Darwinian Controversies*. Cambridge, UK: Cambridge University Press.

*Morpurgo-Tagliabue, Guido (1981). *I processi di Galileo e l'epistemologia*. Rome: Armando.

Newton, Isaac (1999). *The Principia: Mathematical Principles of Natural Philosophy*, trans. and ed. I. Bernard Cohen and Anne Whitman. Berkeley: University of California Press.

Numbers, Ronald L. (1993). *The Creationists*. Berkeley: University of California Press.

*Numbers, Ronald L., Ed. (2009). *Galileo Goes to Jail and Other Myths about Science and Religion*. Cambridge, MA: Harvard University Press.

Ortega y Gasset, José (1956). *En torno a Galileo*. Madrid: Revista de Occidente.

Ortega y Gasset, José (1958). *Man and Crisis*, trans. M. Adams. New York: Norton.

Pagano, Sergio, Ed. (1984). *I documenti del processo di Galileo Galilei*. Vatican City: Pontificia Academia Scientiarum.

Pagano, Sergio, Ed. (2009). *I documenti vaticani del processo di Galileo Galilei (1611–1741)*. Vatican City: Archivio Segreto Vaticano.

Pagano, Sergio (2010). "Il precetto del cardinale Bellarmino a Galileo: Un 'falso'?, con una parentesi sul radio, Madame Curie e i documenti galileiani." *Galilaeana* 7:143–203.

Palmieri, Paolo (2001). "Galileo and the Discovery of the Phases of Venus." *Journal for the History of Astronomy* 32:109–29.

Palmieri, Paolo (2008). "Galileus Deceptus, Non Minime Decepit: A Re-appraisal of One of *Dialogo's* Counter-arguments about Extrusion on a Rotating Earth." *Journal for the History of Astronomy* 39:425–52.

Paschini, Pio (1943). "L'insegnamento di Galileo: non temere la verità." *Studium*, April, 39:94–7.

Paschini, Pio (1964). *Vita e opere di Galileo Galilei*, 2 vols. Vatican City: Pontificia Accademia delle Scienze.

Pastor, Ludwig von (1898–1953). *History of Popes from the Close of the Middle Ages*, 40 vols. St. Louis, MO: Herder.

Pesce, Mauro (1992). "Le redazioni originali della lettera 'copernicana' di G. Galilei a B. Castelli." *Filologia e critica* 17:394–417.

Pesce, Mauro (2008). *L'ermeneutica biblica di Galileo e le due strade della teologia cristiana*. Rome: Edizioni di Storia e Letteratura.

Pitt, Joseph (1992). *Galileo, Human Knowledge, and the Book of Nature*. Dordrecht: Kluwer.

*Popper, Karl R. (1956). "Three Views of Human Knowledge." Reprinted in Karl R. Popper, *Conjectures and Refutations*, pp. 97–119. New York: Harper (1963).

Ptolemy, Claudius (1952). *The Almagest*, trans. R. Gatesby Taliaferro. In *Great Books of the Western World*, vol. 16, pp. 1–478. Chicago: Encyclopedia Britannica.

Ranke, Leopold (1841). *The Ecclesiastical and Political History of the Popes of Rome*, 2 vols, trans. S. Austin. Philadelphia: Lea & Blanchard.

Redondi, Pietro (1987). *Galileo Heretic*, trans. R. Rosenthal. Princeton, NJ: Princeton University Press.

*Reeves, Eileen, and Albert Van Helden, trans. and eds. (2010). *On Sunspots*. Chicago: University of Chicago Press.

Ronchi, Vasco (1958). *Il cannocchiale di Galileo e la scienza del seicento*. Turin: Editore Boringhieri.

Ross, Alexander (1646). *The New Planet No Planet*. London.

Rowland, Wade (2001). *Galileo's Mistake: The Archaeology of a Myth*. Toronto: Thomas Allen.

Rowland, Wade (2003). *Galileo's Mistake: A New Look at the Epic Confrontation between Galileo and the Church*, rev. edn. New York: Arcade.

*Russell, Bertrand (1935). *Religion and Science*. Reprinted, New York: Oxford University Press (1997).

Scaglia, Desiderio (1616). *Prattica per proceder nelle cause del Santo Uffitio*. In Alfonso Mirto, "Un inedito del Seicento sull'Inquisizione," *Nouvelles de la République des Lettres* (1986), no. 1, pp. 99–138.

Segre, Michael (1991). *In the Wake of Galileo*. New Brunswick: Rutgers University Press.

*Segre, Michael (1998). "The Never Ending Galileo Story." In *The Cambridge Companion to Galileo*, ed. Peter Machamer, pp. 388–416. Cambridge, UK: Cambridge University Press.

Shea, William R., and Mariano Artigas (2003). *Galileo in Rome*. Oxford: Oxford University Press.

Smith, A. Mark (1985). "Galileo's Proof for the Earth's Motion from the Movement of Sunspots." *Isis* 76:543–51.

*Speller, Jules (2008). *Galileo's Inquisition Trial Revisited*. Frankfurt: Peter Lang.

Toulmin, Stephen, and June Goodfield (1961). *The Fabric of the Heavens*. New York: Harper.

Van Helden, Albert (1984). "Galileo and the Telescope." In *Novità celesti e crisi del sapere*, ed. Paolo Galluzzi, pp. 149–58. Florence: Giunti Barbera.

*Van Helden, Albert, trans. and Ed. (1989). *Sidereus Nuncius, or The Sidereal Messenger*. Chicago: University of Chicago Press.

Van Helden, Albert (1994). "Telescopes and Authority from Galileo to Cassini." *Osiris*, second series, 9:7–29.

Voltaire (1877–83). *Oeuvres Complètes*, 52 vols., ed. Louis Moland. Paris: Garnier.

*Weidhorn, Manfred (2005). *The Person of the Millennium: The Unique Impact of Galileo on World History*. New York: iUniverse.

*Westman, Robert S. (2011). *The Copernican Question*. Berkeley: University of California Press.

White, Andrew D. (1896). *A History of the Warfare of Science with Theology in Christendom*, 2 vols. New York: Appleton.

Winfield, Nicole (2008). "Good Heavens: Vatican Rehabilitating Galileo." *Washington Post*, December 23. Available at http://www.washingtonpost.com/wp-dyn/content/article/2008/12/23; accessed December 24, 2008.

Wootton, David (2010). *Galileo: Watcher of the Skies*. New Haven, CT: Yale University Press.

NOTES

Chapter 1 Introduction

Avoiding Myths and Muddles

1. Voltaire (1877–83), 12:249; Koestler (1964).
2. Einstein (1954), 271; Hawking (1988), 179; (2002), p. xvii.
3. Gingerich (1982), 133; for further details, documentation, and references, see Finocchiaro (1980), 103–66; (1997), 335–56; (2014a), 259–80.
4. See, for example, Finocchiaro (1980), 180–223, 343–412; (2010), 65–96, 229–50, 277–90; (2014a), 243–80; and Chapters 6 and 10, this volume.
5. See, for example, Finocchiaro (1980), 114–15; (2010), 121–34; (2011); (2014a), 259–64; and Chapters 3 and 10, this volume.
6. See, for example, Finocchiaro (1980), 145–79; (1997), 47–69, 335-56; (2010), pp. xiii–xliii, 243–48, 277–90; (2014a), 259–80; and Chapters 8 and 10, this volume.
7. Biagioli (1993, 2006); Finocchiaro (2013).
8. Cf. Ranke (1841), 2:98–125, especially 116–19; Pastor (1898–1953), 28:271–321; Redondi (1987), 227–32; Miller (2008).
9. Here, my account relies primarily on Masini (1621), and to a lesser extent also on Scaglia (1616), Beretta (1998, 1999), and Mayer (2013, 2014, 2015).
10. Masini (1621), 16–17.
11. Masini (1621), 17–18.
12. Masini (1621), 166–7.
13. Masini (1621), 188.

Chapter 2 When the Earth Stood Still

1. Galilei (1890–1909), 2:205–55; see also Cohen (1960), Kuhn (1957), Lindberg (1992), and Toulmin and Goodfield (1961).
2. This diagram is adapted from Kuhn (1957), 31–6, and Harris and Levey (1975), 883.

Chapter 3 The Copernican Controversy (1543–1609)

1. For more details and references, see Finocchiaro (1980), 8-12; (1985); (1997), 306–8; (2010), 144 n. 33.
2. Cf. Finocchiaro (2002b).

3. For example, in his first book, which he sent to Galileo as a gift, Kepler explained why there were exactly six planets in the Copernican system and why their orbital sizes followed the sequence they did. The explanation was based on the idea that the planetary orbits could be identified with the six spheres that can be circumscribed around the five regular polyhedrons (also called Platonic solids), and on the geometrical fact that there are five and only five such solids. See, for example, Dreyer (1953), 373–6; Bucciantini (2003), 3–22.
4. Galilei (1960), 97; Copernicus (1992), 125–6.
5. Galilei (1890–1909), 2:223.
6. Galilei (1890–1909), 1:304–7.
7. See, respectively, Galilei (2008), 97–9, 209.

Chapter 4 Re-assessing Copernicanism (1609–1616)

1. This diagram is taken from Palmieri (2001), 110.
2. Galilei (2008), 46.
3. Drake (1957), 24; (1983), 14; Van Helden (1989), 31.
4. Galilei (2008), 60.
5. Galilei (1890–1909), 3:46; cf. Galilei 2008, 83. I thank David Wootton for bringing this evidence to my attention, although, if I understand him correctly, he (Wootton 2010) wants to interpret it as somehow strengthening Galileo's commitment to Copernicanism.
6. Galilei (1890–1909), 11:11–12.
7. Galilei (1890–1909), 11:11–12.
8. Galilei (1890–1909), 11:11–12.
9. Galilei 1890–1909, 11:344.
10. In Reeves and Van Helden (2010), 296.
11. Drake (1983), p. xix, 133–5.
12. Cf. Reeves and Van Helden (2010), 258–65.
13. Galilei (1890–1909), 12:34–5.
14. In Finocchiaro (1989), 133.
15. Galilei (2008), 111; Finocchiaro (1989), 88; (2014b), 49.
16. Galilei (2008), 110; Finocchiaro (1989), 88; (2014b), 49.
17. Galilei (2008), 111; Finocchiaro (1989), 88–9; (2014b), 49.
18. Galilei (2008), 112; Finocchiaro (1989), 89; (2014b), 50.
19. Galilei (2008), 113; Finocchiaro (1989), 90; (2014b), 51.
20. Galilei (2008), 114; Finocchiaro (1989), 91; (2014b), 51–2.
21. Cf. Blackwell (1991), 23.
22. Guerrini (2009), 47–70.
23. Foscarini (1615), 19.
24. Foscarini (1615), 29–30.
25. Foscarini (1615), 30–1.
26. Foscarini (1615), 34.

27. Foscarini (1615), 35.
28. Boaga (1990), 194.
29. Galilei (2008), 113; Finocchiaro (1989), 90; (2014b), 51.
30. Galilei (2008), 115; Finocchiaro (1989), 92; (2014b), 53.
31. Galilei (2008), 119; Finocchiaro (1989), 96; (2014b), 56.
32. Galilei (2008), 120; Finocchiaro (1989), 96; (2014b), 57.
33. Galilei (2008), 136; Finocchiaro (1989), 110; (2014b), 70.
34. Galilei (2008), 140; Finocchiaro (1989), 114; (2014b), 73.
35. Galilei (2008), 140; Finocchiaro (1989), 114; (2014b), 73.
36. Augustine, *On the Literal Interpretation of Genesis*, book 2, chapter 9, as translated in Finocchiaro (1989), 95; (2014b), 55; Galilei (2008), 118.
37. Augustine, *On the Literal Interpretation of Genesis*, book 1, chapter 21, as translated in Finocchiaro (1989), 101; (2014b), 61; Galilei (2008), 126.
38. For more details and references, see Finocchiaro (2010), 87–9.

Chapter 5 The Earlier Inquisition Proceedings (1615–1616)

1. Galilei (1890–1909), 12:226–27.
2. The report also stated that some of Galileo's wordings in his critical remarks were excessively but unnecessarily negative and harsh. And here it is important to note that there exist two versions of Galileo's letter to Castelli, one harshly worded and another milder one. The Inquisition consultant examined the harsher one, and judged it essentially unobjectionable, despite some minor flaws in its tone. And Galileo himself had some inklings that the harshly worded version was unnecessarily offensive, and revised it by toning down some of its language. These issues have recently received considerable attention, when a previously unknown manuscript was discovered in the archives of the Royal Society in London; it seems to be the original letter to Castelli in Galileo's own handwriting, with the harsh wordings deleted and corrected into milder ones. See Galilei (1890–1909), 19:305; Finocchiaro (1989), 135–6; Pesce (1992); (2008), 29–53; Camerota, Giudice, and Ricciardo (2018); Abbott (2018); and Anonymous (2018).
3. Finocchiaro (1989), 147; (2014b), 102.
4. Finocchiaro (1989), 147–8; (2014b), 102–3; Galilei (2008), 175–6.
5. For this particular thesis, especially incisive is the analysis in Speller (2008), 96–8.
6. Finocchiaro (1989), 148–50; (2014b), 103–4; Galilei (2008), 176–8.
7. Finocchiaro (1989), 200–2.
8. Finocchiaro (1989), 148; (2014b), 103.
9. Finocchiaro (1989), 153; (2014b), 105; Galilei (2008), 178.
10. On this particular point, see Kelly (2015); (2016).

Chapter 6 The *Dialogue on the Two Chief World Systems* (1632)

1. As previously noted, for more details and references on this argument, see Finocchiaro (1980), 8-12; (1985); (1997), 306-8; (2010), 144 n. 33.

2. For documentation of this interpretation and criticism of alternatives, see Finocchiaro (1980), 16–18, 76–8; Drake (1986); MacLachlan (1990); Pitt (1992), 78–109; and Chapter 4, this volume.
3. Galilei (1997), 104.
4. Finocchiaro (1989), 184.
5. This diagram is taken from Smith (1985), 544; cf. Galilei (1953), 348.
6. This diagram is adapted from Smith (1985), 545.
7. Galilei (1997), 235, 237; cf. Galilei (1953), 328, 335.
8. This is adapted, with some obvious additions, from Galileo's *Dialogue* (Galilei (1953), 426; (1997), 291). I thank Mark Attorri for suggesting these additions, as well as for many other valuable comments.
9. For example, in Galilei (1953), 352.
10. Galilei (1953), 60.
11. Galilei (1890–1909), 7:292–3; cf. Galilei (1953), 268; Finocchiaro (1997), 365.
12. Galilei (1997), 147; cf. Galilei (1953), 127.
13. Galilei (1997), 233; cf. Galilei (1890–1909), 7:355. Here, Galileo's meaning will be completely missed if one relies on Drake's English translation (Galilei (1953), 327) or that by Davie (Galilei (2012), 307); even Strauss, who gets the German translation right (Galilei (1982), 342), misses the essential point in his comment (Galilei (1982), 551 n. 34); cf. Finocchiaro (1980), 230–1, 244.
14. Galilei (1997), 284; cf. Galilei (1953), 420.

Chapter 7 The Inquisition Trial (1632–1633)

1. Here, I am following in part the interpretation in Morpurgo-Tagliabue (1981).
2. Finocchiaro (1989), 218–22; (2014b), 119–22; Galilei (2008), 272–6.
3. For an elaboration of this conjecture, see Beltrán Marí (2006), 496–528.
4. For an elaboration of this conjecture, see Speller (2008), 143–60, 375–96.
5. Seghizzi, who had been the commissary in 1616, had died in 1625.
6. Finocchiaro (2005), 247; (2014b), 132–3.
7. Masini (1621), 120-51; Finocchiaro (2009).
8. Finocchiaro (1989), 287–91; (2014b), 134–8; Galilei (2008), 288–93.
9. Finocchiaro (1989), 291; (2014b), 138; Galilei (2008), 292.
10. For details, see Finocchiaro (2009), 68–74.
11. Finocchiaro (1989), 292–3; (2014b), 138–9; Galilei (2008), 293–4.
12. Finocchiaro (1989), 292; (2014b), 139; Galilei (2008), 293.
13. Finocchiaro (1989), 292; (2014b), 139; Galilei (2008), 293.
14. For example, Genovesi (1966), 268.

Chapter 8 Becoming a Cultural Icon (1616–2016)

1. See Mayaud (1997), 52.
2. For more details and documentation, see Mayaud (1997); Heilbron (1999; 2005); Finocchiaro (2005), 126–53, 193–221.

3. For more details and documentation, see Heilbron (1999), 207–11; (2010), 362–5; McMullin (2005), 323–60; Finocchiaro (2005); (2010), pp. xx–xliii, 229–338.
4. Paschini (1964).
5. Leo XIII (1893), paragraph 18, p. 334. Cf. Augustine, *On the Literal Interpretation of Genesis*, book 1, Chapter 21, translated in Finocchiaro (1989), 101; (2014b), 61; Galilei (2008), 126.
6. Leo XIII (1893), paragraph 18, p. 334. Cf. Augustine, *On the Literal Interpretation of Genesis*, book 2, Chapter 9, translated in Finocchiaro (1989), 95; (2014b), 55; Galilei (2008), 118.
7. For more details and documentation, see Acloque (1982); Heilbron (1999), 208–9, 235–8, 300–2; (2005); Finocchiaro (2010), pp. xxi–xxii.
8. Galilei (1890–1909), 7:259–60; (1953), 233–4.
9. For more details and documentation, see Finocchiaro (2010), pp. xx–xxxi, 155–314.
10. Galluzzi (2000), 539.
11. For details, see Koyré (1955); Galluzzi (1977); Heilbron (2010), 359–60.
12. John Paul II (1992), section 5, paragraphs 4–5.
13. John Paul II (1992), section 7, paragraph 2.
14. Mallet du Pan (1784), 122; cf. Finocchiaro (2002a).
15. Galilei (1890–1909), 15:55–6.
16. Paschini (1943), 97; cf. Finocchiaro (2005), 280–4. In one of his unpublished writings dated 1980, Stillman Drake made a similar point; see DiCanzio (1996), 309.
17. Duhem (1908), 136; cf. Duhem (1969), 113.
18. See Morpurgo-Tagliabue (1981); Finocchiaro (2010), Chapters 3, 9, 11, and 12.
19. Rowland (2001; 2003).
20. For more details and documentation, see Finocchiaro (2005), 295–317; (2010), pp. xxxi–xxxviii, 300–1.
21. Feyerabend (1985), 164.
22. Feyerabend (1988), 129; (1993), 125.
23. Benedict XVI (2006).
24. See http://w2.vatican.va/content/benedict-xvi/en/angelus/2008/documents/hf_ben-xvi_ang_20081221.html, consulted on January 29, 2019.
25. Winfield (2008).
26. Bucciantini, Camerota, and Giudice (2011).
27. For more details and documentation, see Pagano (1984; 2009; 2010); Finocchiaro (2016; 2018); see also Chapter 5, this volume.
28. Obviously, the 2016 anniversary is merely the last episode discussed here, and not the end of the story. In fact, the August 2018 discovery of Galileo's original letter to Castelli promises to rekindle some issues, as suggested from the attention given to it by the journal *Nature*. See Abbott (2018); Anonymous (2018); Camerota, Giudice, and Ricciardo (2018); and Chapter 5, note 2, this volume.
29. See, respectively, http://cmrs.ucla.edu/event/2863/; http://origins.osu.edu/milestones/february-2016-400-years-ago-catholic-church-prohibited-copernicanism; and http://www1.udel.edu/udaily/2016/feb/galileo-020416.html.

Chapter 9 Religion vs. Science?

1. Draper (1874); White (1896). For some criticism, see Lindberg and Numbers (1987); Brooke (1991), 34–7.
2. Voltaire (1877–83), 12:249; Russell (1935), 31–43; Einstein (1953), 7; Popper (1956). For more details and references, see Finocchiaro (2002b); (2005), 115–25; (2009); (2010), 293–6; (2014a), 311–14.
3. John Paul II (1979), section 6, paragraph 1.
4. John Paul II (1979), section 6, paragraph 2.
5. John Paul II (1979), section 7, paragraph 1.
6. On the non-monolithic character of the Catholic Church, see also Segre (1991), 30; Feldhay (1995); Lindberg (2003), 58; Speller (2008); and Mayer (2012), 3.
7. For similar views and additional support, see Kuhn (1977); Moore (1979), 80–103; and Numbers (1993), pp. xiv–xv.
8. Campanella (1994).
9. The French scholar Michel-Pierre Lerner deserves credit for having stressed this; see Lerner (2001), pp. xcv–c, 1–2.
10. Campanella (1622), 30; cf. Campanella (1994), 79, where the English translation misses this particular nuance.
11. Galilei (1890–1909), 12:287.
12. Campanella (1622), 50; cf. Campanella (1994), 110.
13. Campanella (1994), 65.
14. Campanella (1994), 54.
15. Campanella (1994), 65–6.
16. Campanella (1994), 54.
17. Campanella (1994), 69.
18. Campanella (1994), 57.
19. Campanella (1994), 71.
20. Campanella (1994), 74.
21. Campanella (1994), 74.
22. Campanella (1994), 76.
23. Campanella (1994), 97.
24. Campanella (1994), 98.
25. Campanella (1994), 99.
26. Campanella (1994), 122–3.
27. Cf. Finocchiaro (2005), 72–6.
28. Milton (1644), 537–8; I have modernized the spelling.
29. Voltaire (1877–83), 12:249.
30. Libri (1841), 46–7.
31. Einstein (1953), p. vii.
32. Ross (1646), 9.
33. Mallet du Pan (1784), 122.
34. Brewster (1841), 93–5.
35. Duhem (1908), 136; (1969), 113; Feyerabend (1988), 129; and cf. Chapter 8, this volume.

36. For a clarification of the difference between these two buildings, often confused even by scholars, see Shea and Artigas (2003), 30, 74, 106–7, 134–5, 179–80, 195.
37. For more details, see Berggren and Sjöstedt (1996), 19–20, 145–7.
38. "Epigrafi ed Offese," *L'Osservatore Romano*, April 23, 1887.
39. For a general account of myths, see Carvalho Neto (1971); for partial applications to the relationship between science and religion, see Livingstone (2007), Numbers (2009); for partial applications to the Galileo affair, see Benítez (1999), 85–110; Finocchiaro (2002a; 2009); Lessl (1999); Segre (1998).
40. Cf. Brown (1979).
41. Beltrán Marí (1998); Benítez (1999), 85–110.

Chapter 10 A Model of Critical Thinking?

1. For one of the first elaborations of this thesis, see Butterfield (1949).
2. Newton (1999), 424; Einstein (1953), p. xiii; Hawking (1988), 179; (2002), p. xvii; Hume (1851–60), 4:521–7; (1948), 24–5; Kant (1965), 20–2; Ortega y Gasset (1956); (1958), 9–29. Cf. Finocchiaro (1980), pp. xv–xx, 92–7; (2014a), 243–5, 314–16.
3. For more details about the telescope controversy, see for example Feyerabend (1975; 1988); Ronchi (1958); Van Helden (1984; 1994).
4. For more details, see, for example, Hill (1984); Finocchiaro (2010), 97–120; Palmieri (2008).
5. Galilei (1997), 173; cf. Galilei (1953), 190.
6. Aristotle, *On Sophistical Refutations*, 167a21.
7. Hill (1984), 110–12. The truth of the matter is that Galileo was presumably embellishing a version formulated by Copernicus (*On the Revolutions*, I, 7; (1992), 14–15). The latter was attributing the argument to Ptolemy. But the relevant passage from Ptolemy (*Almagest*, I, 7; (1952), 10–12) is insufficiently clear and explicit: it is hard to determine whether he indeed had the extrusion objection in mind or was talking about terrestrial bodies being left behind if the whole Earth were moving. For further clarifications about this aspect of the question, see Palmieri (2008).
8. Galilei (1953), 373–9, 389–97; cf. Finocchiaro (2014a), 201–9.
9. Galilei (1997), 234; (2008), 242.
10. Galilei (1997), 235; (2008), 242.
11. Galilei (1997), 147–8.
12. In Finocchiaro (1989), 85; (2014b), 93–4.
13. Finocchiaro (1989), 278; (2014b), 129; Galilei (2008), 283.

Chapter 11 Some Final Thoughts

1. See, for example, Brooke (1991), 5, 8–10, 33, 42, 50–1; Brooke and Cantor (1998), pp. xi, 21, 66; Lightman (2001); Lindberg and Numbers (1986), 6, 10, 14; (2003), 3.

ARTWORK ACKNOWLEDGMENTS

Figure 1, The Metropolitan Museum of Art, New York, Gift of Phyllis D. Massar, 2002; Figure 2, Collegium Maius, Krakow. Photo © Image Asset Management/World History Archive/age fotostock; Figure 9, Museo delle Scienze, Florence. Photo © akg-images; Figures 10 and 12, Library of Congress, Rare Book and Special Collections Division; Figure 13, The Metropolitan Museum of Art, New York, The Elisha Whittelsey Collection, The Elisha Whittelsey Fund, 1951; Figure 14, Musei Capitolini Roma. Photo © Archivart/ Alamy/age fotostock; Figures 15 and 16, A. Mark Smith, "Galileo's Proof for the Earth's Motion from the Movement of Sunspots," *Isis* 76(1985): 543–51, at pp. 544 and 545, published by the University of Chicago Press; copyright © 1985 by the History of Science Society, Inc.; Figure 18, Library of Congress, Rare Book and Special Collections Division; Figure 19, iStock.com/photooiasson; Figure 20, Lasse Ansaharju/Shutterstock.com.

INDEX